The Impact of Climate Change on Sub-Saharan Africa

AF764823

STUDIES IN SUB-SAHARAN AFRICA

Edited by Support Africa International/Support Africa Stiftung

Represented by Baldur Pfeiffer

Vol. 6

PL ACADEMIC RESEARCH

Chukwuemeka Christopher Opara /
Bernard Palmer Kfuban Yerima (eds.)

The Impact of Climate Change on Sub-Saharan Africa

Case Studies in Cameroon, Nigeria and Uganda

PL ACADEMIC RESEARCH

Bibliographic Information published by the Deutsche Nationalbibliothek
The Deutsche Nationalbibliothek lists this publication in the Deutsche Nationalbibliografie; detailed bibliographic data is available in the internet at http://dnb.d-nb.de.

Library of Congress Cataloging-in-Publication Data
The impact of climate change on Sub-Saharan Africa : case studies in Cameroon, Nigeria and Uganda / Chukwuemeka Christopher Opara, Bernard Palmer Kfuban Yerima (eds.). – First edition.
pages cm. – (Studies in sub-Saharan Africa ; vol. 6)
ISBN 978-3-631-65347-0
1. Climatic changes–Africa, Sub-Saharan. 2. Climatic changes–Cameroon. 3. Climatic changes–Nigeria. 4. Climatic changes–Uganda. 5. Climate change mitigation–Africa, Sub-Saharan. 6. Climate change mitigation–Cameroon. 7. Climate change mitigation–Nigeria. 8. Climate change mitigation–Uganda. I. Opara, Chukwuemeka Christopher, 1963- II. Yerima, Bernard P. K. (Bernard Palmer Kfuban), 1952- III. Series: Studies in sub-Saharan Africa ; v. 6.
QC903.2.A357I48 2014
363.738'7420967–dc23
2014033600

This research and publication was sponsored by the generous support of the Grebe Group, Weilburg, Germany.

ISSN 1614-001X
ISBN 978-3-631-65347-0 (Print)
E-ISBN 978-3-653-04584-0 (E-Book)
DOI 10.3726/978-3-653-04584-0

© Peter Lang GmbH
Internationaler Verlag der Wissenschaften
Frankfurt am Main 2014
All rights reserved.
PL Academic Research is an Imprint of Peter Lang GmbH.

Peter Lang – Frankfurt am Main · Bern · Bruxelles · New York · Oxford · Warszawa · Wien

All parts of this publication are protected by copyright. Any utilisation outside the strict limits of the copyright law, without the permission of the publisher, is forbidden and liable to prosecution. This applies in particular to reproductions, translations, microfilming, and storage and processing in electronic retrieval systems.

This publication has been peer reviewed.

www.peterlang.com

Foreword to the Series

The research findings published in this series, Studies in Sub-Saharan Africa, have been promoted by SUPPORT AFRICA INTERNATIONAL of the SUPPORT AFRICA FOUNDATION, Nassau, Germany.

The articles and contributions published in this book are a selection of research papers submitted to the annual calls for the *Research Award for Ecology* of SUPPORT AFRICA INTERNATIONAL. Besides ecology there are also calls for research awards in *Health/Medicine* and in *Sustainable Agriculture and Biotechnology*. The results of these studies are likewise published in this series by the organization mentioned above.

The findings reflect the present state of research and development projects in the respective scientific fields at Sub-Saharan universities. The authors are independent scientists taking responsibilities for their studies published.

Publishing this series is to encourage institutional research and development of universities in Sub-Sahara Africa. It appeals to members of academic departments, providing them with a platform to voice their concerns. Though this series is open to a world-wide readership, it primarily aims at reaching the scientific world in Sub-Sahara Africa.

Baldur Ed. Pfeiffer
Editor-in-Chief

Uche C. Amalu
Co-Editor-in-Chief

Table of Contents

Uche C. Amalu
University of Calabar, Calabar, Nigeria

Foreword to this Volume

Climate change, one of the most recent phrases in global environment vocabulary, is also one of the greatest transcendental challenges of this century.

Often used interchangeably with global warming, climate change is the slow change in the composition of the global atmosphere, which is caused directly and indirectly by various human activities in addition to natural climate variability over time. Just as global warming aptly describes the process of warming up of the globe, climate change also describes changes in the atmosphere over a period of time. However, the definitions and descriptions of climate change have become an evolving concept, describing changes in the sum total, long-term averages of the globe. In a strange way, and in very diverse ways, climate change affects everybody in the world, but in the most fundamental way, climate change will bring change to agriculture wherever it is practiced around the globe.

During the 1980's, scientists, governments and the public became more concerned about the possibilities that the world may be getting warmer. Ever since then various studies and assessments have been carried out by researchers worldwide. There can no longer be any doubt – the earth's climate is changing, and some nations will likely experience more adverse effects of climate change than other nations. Some nations may benefit more than others and poorer nations are generally more vulnerable to the consequences of global warming. Indeed the poorer nations tend to be more dependent on climate-sensitive sectors, such as agriculture, forestry and natural resources, and they lack the economic resources to buffer themselves against the changes that global warming may bring. Considering the foregoing factors and because of widespread poverty, food and nutrition insecurity, both of which limit adaptation capabilities, Africa, but particularly the Sub-Saharan sub-region, is the continent most vulnerable to the impacts of projected changes in climate. While the specific consequences resulting from global warming will continue to be debated, areas of concern include rising sea levels, disruption of the hydrologic cycle, worsening human-health effects, changing forests and natural areas and challenges to agriculture and food supply, factors which are not only relevant to Africa as a whole, but also very determinate to the continued survival of the sub-regions teeming populations.

The volume entitled **The Impact of Climate Change on Sub-Sahara Africa: Case Studies in Cameroon, Nigeria and Uganda**, as the other books in our series before, should present the scenarios in countries of Sub-Saharan Africa and from African scientists themselves, but this volume emphasizes results of various studies and assessments made by scientists from Cameroon, Nigeria and Uganda. The scenarios in the aforesaid countries aptly demonstrate impacts and manifest experiences across the Sub-Saharan sub-region in Africa. Furthermore, majority of the thirteen articles come from Nigerian authors, who also derived their examples and references from Nigerian perspectives, while the others come from authors in Cameroon and Uganda.

The volume is divided in three sections: section one, "Global climate change impacts on Sub-Saharan Africa" present articles on causes, observations, mitigation and adaptation strategies, as well as issues in the Nigerian shorelines. The second section discusses several measures and strategies for facing climate change in the sub-region. The emerging contributions of Information and Communication Technologies (ICTs) and some measures for the protection and sustainability of plant and animal species and other forms of biodiversity were given special attention. Beyond recommendations, actions of individuals, organizations and governments (especially in the Yongka Nkwen-Bamenda, Northwest region, Cameroon) in combating climate change are discussed in section three. The articles are written in an easy-to read and understand styles and hereby recommended for reading by the discerning public, teachers and researchers in colleges and universities, as well as by decision-takers and policymakers in government.

I. Global climate change impacts on Sub-Saharan Africa

Chukwuemeka Christopher Opara
Federal University of Technology, Owerri, Nigeria

Climate change phenomenon: Causes, consequences, mitigation and some adaptation strategies in Nigeria

Introduction

The term climate change has been given different definitions by various researchers. For instance, Cook (2005) defined climate change as a change in the statistical distribution of weather over periods of time that range from decades to millions of years. Thus, it could be any long-term significant change in the average weather or in the distribution of weather events around an average (for example, greater or fewer extreme weather events) in a given region. In the context of environmental policy, climate change refers to changes in modern climate giving rise to global warming (IPCC, 2007). The International Council for Science (ICSU, 2007) indicated that climate change is one of the "global environmental changes (GEC)" and global environmental change refers to the set of transformation of land, oceans and atmosphere driven by inter-woven, socio-economic and natural processes. The Framework Convention on Climate Change (FCCC) in its Article 1 and cited in Green (2008) referred to climate change as a change of climate, which is attributed directly or indirectly to human activity that alters the composition of the global atmosphere and, which is in addition to natural variability observed over comparable time periods. Climate changes reflect variations within the earth's atmosphere processes in other parts of the earth, such as oceans and ice caps, and the effect of human activities. Most forms of internal variability in the climate system could be recognized as a form of hysteresis, indicating that the current state of climate reflects not only the inputs but also the history of how it got to a given point or place. For instance, a decade of dry conditions may cause lakes to shrink. Indeed, climate change could be a self-perpetuating process, because different aspects of the environment respond in different ways to the fluctuations that inevitably occur.

The problems associated with climate changes have, however, been reported to include atmospheric temperature increases or decreases, which could give rise to changes in rainfall distribution and growing season length and greater frequency and severity of storms (Brady and Weil, 2005; Cunningham *et al.*, 2005); sea level increase or decrease (Brady and Weil, 2005; Boesch *et al.*, 2000; Cunningham *et al.*, 2005; Green, 2008; Woodroffe and Horton, 2005) aggravating weather extremes such as flooding, drought and desertification which in turn result in soil quality breakdown, poverty food insecurity, social tensions and migration of people and breakdown in human health, water resources and natural resources – based livelihoods (ICSU, 2007) especially in Sub-Saharan Africa. The issues and problems associated with natural (forest products and land) and human (indigenous knowledge, agricultural skills/training and water management) resources and their current or potential relationship with climate have continued to receive a high level of national and international attention. Thus, it should come as no surprise that climate variability and the potential consequences of future climate change should be of considerable interests to stakeholders, policy makers and the general public, especially as they may act as an additional stressor to land and forest resources. However, regardless of the actions taken to mitigate or otherwise address the causes and consequences of climate change in the Sub-Saharan Africa, one of the most important responses to climate change would be to deliver timely information and current useful or relevant scientific findings to decision makers and the public. According to Boesch *et al.* (2000), to deliver such timely information and relevant scientific findings, the best available scientific information on the patterns, projections and consequences of both climate variability and change should be widely used, together with an understanding of associated uncertainties and should lead to informed policy making that protects the environment and ensures a sustainable future for the nation.

In this study, a 30 year rainfall data obtained from Owerri, south-eastern Nigeria, was used to show the degree or extent of reliability or dependability and perhaps the existence of climate change phenomenon in the region. The ICSU (2007) report has shown that the most crucial climate change issue for Africa's (especially Sub-Saharan Africa's) decision-makers at all levels, from the household to the African Union, is rainfall, and in particular its variability at all time-scales, hence its use in this study even though it is also associated with the greatest uncertainty in climate change projections. The use of rainfall data in this study was also based on the fact that rainfall measurements, analysis, modeling and evaluation have been shown to be inadequate across much of the African regions, especially Sub-Saharan Africa. Thus, this is also an attempt to promote the sharing of rainfall information with other regions of the world. In general, however, there is still a dearth of information on evidence of causes and effects of climate change in Sub-Saharan Africa, especially Nigeria's south-eastern region, thereby necessitating the sensitization of both the government and public to the existence of climate change.

This study reviews the climate change phenomena: causes, consequences and mitigation and adaptation strategies in Nigeria, with emphasis on south-eastern Nigeria. This has become necessary, especially now that so many people still have the impression that the climate change phenomenon is still nebulous, perhaps due to lack of appreciable knowledge of climate changes. Knowledge science is a veritable tool for proper understanding of climate change. Secondly, the public may admit observing changes around them, but would attribute such changes to factors other than climate change, such as sinful generation, wrath of God, signs of the end time or end of life, etc. In general not everyone believes that the global temperatures are increasing because of the burning of coal and other fossil fuels or that the humans have the ability to change the global temperature or cause climate change (Glantz, 2005; Green, 2008). In other words, there is so much disagreement about whether human activities cause global climate change perhaps because the weather and climate are complex and highly variable (Cunningham *et al.*, 2005).

The factors that can shape climate are called climate forcings. These include such processes as variations in solar radiation, deviations in the earth's orbit, mountain building and continental drift that are natural phenomena and changes in greenhouse gasses that could either be naturally or anthropogenically induced. However, ICSU (2007) had indicated that concurrent with the modern phenomenon of globalization (the growing and accelerate interconnectedness of the world) and, to some extent, as a consequence of it, people have begun to induce planetary – scale changes in the earth's life-support system. The growing demographic pressure and increased demand for food, fiber and water have dramatically transformed the land surface from quasi-natural cover to cultivated lands. The carbon dioxide liberated during this process combines with emissions from the burning of fossil fuels to change the composition and radiant energy balance of the atmosphere, leading to a warming of the earth's surface, changes in the distribution and quantity of rainfall; and a rise in the mean sea level, (McCarthy *et al.*, 2001; Boko *et al.*, 2007). Unsustainable harvest pressures and loss of habitat for wild species have precipitated loss of biodiversity and, at the same time, a small number of invasive species are increasing their dominance. Thus human activities now match (and often exceed) the natural forces that regulate the earth system. Particles emitted by anthropogenic activities alter the energy balance of the planet, and have adverse effects on human health (Charlson *et al.*, 1992). Anthropogenic changes of the structure, composition, or function of large natural biogeochemical and ecological systems entail changes in the complex array of forcings and feedbacks that characterize the internal dynamics of the earth system (Steflen *et al.*, 2004). These planetary-scale environmental changes are of great significance, because they diminish the capacity of the earth's natural environment to supply and replenish resources, and to absorb and recycle the waste products from people's activities. These perturbations are exacerbated by global climate change, and together, they compromise livelihoods by constraining resources in poorer communities, for example, and by affecting tourism and the range of other goods and services that ecosystems provide (MEA, 2005). Human societies are themselves changing, from rural to urban and from locally-focused to globally-connected.

Causes of climate change

The atmosphere is never static, but always in constant turmoil. There are important interactions within the atmosphere resulting from or causing such changes. These are called feedback mechanisms since there are no simple one-way causes and effects processes as effects often rebound to alter their causes (Ayoade, 1993). Thus, the changes in the atmosphere may be internally induced within the earth-atmosphere system or externally induced by extra-terrestrial factors. In the lowest 10 to 12 km, a layer called "the troposphere, the air moves or circulates ceaselessly in great vertical and horizontal convection currents, flowing and swirling and continuously redistributing heat and moisture from one part of the globe to another (Cunningham *et al.*, 2005). The composition and behavior of the troposphere and other layers thus control the weather (daily temperature and moisture conditions in a place) and the climate. Thus, a change in climate implies a change in the general circulation of the atmosphere on which climate ultimately depends (Ayoade, 1993). Therefore, that the atmosphere is not static also indicates that the earth's climate is dynamic and always changing through a natural cycle. However, what the world is worried about is that the changes that are occurring today have been speeded up following man's activities.

Natural causes

There are a number of natural factors responsible for climate changes, which include the movement of tectonic plates. The motion of tectonic plates tends to reconfigure global land and ocean areas and generate topography. This can affect both global and local patterns of climate and atmosphere-ocean circulation (Forest *et al.*, 1999).

The position of the continents determines the geometry of the ocean and therefore influences patterns of ocean circulation. The locations of the seas are important in controlling the transfer of heat and moisture across the globe and therefore in determining global climate. For instance, the proximity of the southern part of Nigeria to the Atlantic Ocean would explain why the people living therein experience more rainfall than those living in northern Nigeria, due to the movement of southwest trade winds, which carry moisture from the Atlantic Ocean into the coastal region and hinterlands of southern Nigeria.

Topography can also influence climate. The existence of mountains can cause orographic precipitation. Humidity generally decreases and diurnal temperature swings generally increase with increasing elevation. Mean temperatures and the length of the growing season also decrease with increasing elevation. The yearly temperature variations are generally lower in coastal areas than they are inland, because of the stabilizing effect of oceans on temperature (Wikipedia, 2009).

Continental drift

It is believed that South American and African continents were formed following a gradual drifting apart of the land mass millions of years back; this drift had impacts on the climate, because it changed the physical features of the land mass, their position and the position of water bodies. The separation of the land masses changed the flow of ocean currents and winds, which affected the climate (Wikipedia, 2009). The oceans are a major component of the climate system. They cover about 71% of the earth and absorb about twice as much of the sun's radiation as the atmosphere or the land surface. Ocean currents move vast amounts of heat across the planet – roughly the same amount as the atmosphere does. But the oceans are surrounded by land masses, so heat transport through the water is through channels. Much of the heat that escapes from the oceans is in the form of water vapor (the most abundant greenhouse gas on earth). Yet, water vapor also contributes to the formation of clouds, which shade the earth's surface and have a net cooling effect.

Solar Output

The sun is the predominant source of energy input to the earth. It supplies the earth with an enormous amount of energy, except that the energy is not evenly distributed over the globe (Cunningham, *et al.*, 2005). Both long- and short-term variations in solar intensity are known to affect global climate. Of the solar energy that reaches the outer atmosphere, about one quarter is reflected by clouds and atmospheric gases and another quarter is absorbed by carbon dioxide, water vapor, ozone, methane and a few other gases. The absorbed energy would in turn warm the atmosphere slightly. About half of the incoming solar radiation (insolation) reaches the earth's surface. Some of this energy is reflected by bright surfaces such as snow, ice and sand. The rest is absorbed by the earth's surface and water. Surfaces that reflect energy have high "albedo" (reflectivity or a surface's reflective properties). Surfaces that absorb energy have low albedo and generally appear dark. Such surfaces include black soils, dark green vegetation, etc. Absorbed energy heats the absorbing surface, evaporates water or provides energy for photosynthesis in plants. Following the second law of thermodynamics, absorbed energy is gradually remitted as lower quality heat energy. The change in energy quality is very important because the atmosphere selectively absorbs longer wavelengths. Most solar energy comes in the form of intense, high-energy, light or near infrared wavelengths. This short-wavelength energy passes relatively easily through the atmosphere to reach the earth's surface. Energy re-released from the earth's warmed surface (terrestrial energy) has lower-intensity and longer-wavelength energy in the far-infrared part of the spectrum. Atmospheric gases, especially carbon dioxide and water vapor, absorb much of this long-wavelength energy, re-releasing it in the lower atmosphere and letting it leak out to space only slowly. This terrestrial energy provides most of the heat in the lower atmosphere. If the atmosphere were as transparent to infrared radiation as it is to visible light, the earth's average surface temperature would be about 33^0C colder than it is now. This phenomenon is called the "greenhouse effect" because the atmosphere, loosely comparable to the glass of a greenhouse, transmits sunlight, while trapping heat inside (Cunningham *et al.*, 2005). Thus, the "greenhouse effect" is the entrapment of heat (the outgoing long-wave radiation) by upper atmospheric gases called greenhouse gases (such as carbon dioxide, water vapor, methane, chlorofluorocarbons, ozone and nitrous oxide), while allowing the short wavelength solar radiation in, to result in the earth becoming much warmer (hence global warming) than it would otherwise be (Brady and Weil, 2005). Some soils can accumulate large quantities of carbon as soil organic matter (OM), thereby, having a major impact on such global changes as the greenhouse effects. Through aerobic decomposition, the carbon in plant biomass and soil OM is eventually converted back into CO_2 and returned to the atmosphere.

Contributions to greenhouse gases

Methane (CH_4) contribution to the greenhouse effect is nearly half as great as that from CO_2 in trapping outgoing radiation. The level of CH_4 is rising at about 0.6% per annum. In the year 2000, there was about 1.8 ppm CH_4 in the atmosphere, more than double the pre-industrial level. Soils serve as both a source and a sink for CH_4.

Biological soil processes account for much of the methane emitted into the atmosphere. About 25% of all methane emissions are said to come from domesticated animals such as dairy cows, goats, pigs, buffaloes, camels, horses and sheep. These animals produce CH_4 during the cud-chewing process (Cunningham *et al.*, 2005; Wikipedia, 2009). Also, significant quantities of CH_4 are produced by the anaerobic decomposition of cellulose in the guts of termites living in well aerated soils and garbage buried deep in landfills (Brady and Weil, 2005). Methane is also released from rice or paddy fields that are flooded with water due to anaerobic conditions, which favor methane producing bacteria and other organisms that decompose OM in the soil to produce CH_4. Furthermore, CH_4 is emitted during the process of oil drilling, coal mining and from leaking gas pipe lines (Cunningham *et al.*, 2005; Wikipedia, 2009).

Nitrous oxide (N_2O) is produced by fertilizer applications, de-nitrification processes, burning of OM and leguminous plants such as beans and pulses (that add nitrogen to the soil) (Cunningham *et al.*, 2005; Wikipedia 2005).

It is worthy to note that of all the greenhouse gases, the chlorofluorocarbons (CFCs) are the only gases that are exclusively of industrial origin (Brady and Weil, 2005). In general, however, reports have shown that since the pre-industrial times, CO_2, CH_4 and N_2O have increased by 31%, 151% and 17%, respectively. The relative contribution of some of the greenhouse gases to the global greenhouse effects or global warming is in the following decreasing order: CO_2 (64%), CH_4 (19%), CFCs (11%) and N_2O (6%) (Cunningham *et al.*, 2005).

Orbital variations

Slight variations in the earth's orbit lead to changes in the amount of sunlight reaching the earth's surface and how it is distributed across the globe. The former is similar to solar variations in that there is a change to the power input from the sun to the earth system. The latter is due to how the orbital variations affect when and where sunlight is received by the earth. The three types of orbital variations are "variation in the earth's eccentricity", changes in the tilt angle of the earth's axis of rotation" and 'precision of the earth's axis". Combined together, these processes produce Milankovitcti cycles, which have a large impact on the climate (Cunningham *et al.*, 2005; Wikipedia, 2009) and are notable for their correlation with the advance and retreat of the Sahara (Milankovitch cycles and Glaciations) (http://www.homepage.mortane.edu). Thus, Milankovitch cycles are periodic shifts in the earth's orbit and tilt (Cunningham *et al.*, 2005).

The earth makes one full orbit around the sun each year. It is tilted at an angle of 23.5^0 to the perpendicular plane of its orbital path (Wikipedia, 2009). For one half of the year when it is warm (summer), the northern hemisphere tilts towards the sun. In the other half when it is cold (winter), the earth is tilted away from the sun. In the absence of tilt, no seasons would be experienced.

The Earth's orbit is somewhat elliptical, which means that the distance between the earth and the sun varies over the course of a year. The earth's axis is not quite constant but the axis moves at the rate of a little more than a half-degree each century. Thus, this gradual change in the direction of the earth's axis, called precession, is responsible for changes in climate (Wikipedia, 2009).

In general, these variations following shifts or changes in the earth's tilt and orbit would change the distribution and intensity of sunlight reaching the earth's surface and, consequently, global climate (Cunningham *et al.*, 2005).Volcanism is a process of conveying material from the crust and mantle of the earth to its surface in form of gases (sulphur dioxide, SO_2 and water vapor), ash or molten lava or magma (Cunningham *et al.*, 2005). Eruptions, large enough to affect the climate occur on average several times per century and cause cooling (by partially blocking the transmission of solar radiation to the earth's surface). In this case, the gases and dust particles partially block the incoming rays of the sun leading to cooling. SO_2 combines with water to form tiny droplets of sulphuric acid (H_2SO_4). These droplets are efficient reflectors of sunlight and screen the ground from some of the energy that it would ordinarily receive from the sun. Winds in the upper levels of the atmosphere called the stratosphere carry the aerosols rapidly around the globe. The eruption of mountain Pinatubo in the Philippines in 1991 ejected enough dust, aerosols and gases into the atmosphere to cool the climate by at least 1^0 C (Cunningham *et al.*, 2005).

Anthropogenic factors

Anthropogenic factors are the human activities that change the environment. Even though there is so much disagreement about whether human activities contribute to global climate change, perhaps because weather and climate are complex and highly variable, many scientists regard anthropogenic (human-caused) global climate change to be the most important environmental issue of our times (Cunningham, *et al.*, 2005). Humans have contributed to global warming through the following ways:

Fossil fuel burning

The Industrial Revolution that began in the 19th century saw a sharp increase or large-scale burning of fossil fuels (coal, oil and natural gas) giving rise to high release or emission of CO_2 (a greenhouse gas) into the atmosphere. The emitted CO_2 gas, which is the most important cause of anthropogenic climate change (Cunningham *et al.*, 2005) and undoubtedly the most important greenhouse gas in the atmosphere (Wikipedia, 2009) would entrap the heat energy from solar radiation, to increase the air or atmospheric temperature, thereby resulting in the earth becoming much warmer than it would otherwise be (Glantz, 2005; Cunningham *et al.*, 2005; Brady and Weil, 2005; Wikipedia, 2009). Fossil fuels supply most of the energy needed to run vehicles, generate electricity for industrial activities and households. Electricity is the main source of power in urban areas. All gadgets run on electricity generated mainly from power plants. Those thermal power plants are run with fossil fuels (mostly coal) and are responsible for the emission of huge amounts of greenhouse gases and other pollutants. In the kitchen, stoves make use of kerosene and natural gas that is fossil fuel used in cooking. CO_2 is liberated in the cooking process. Cars, buses and trucks are the ways by which goods and people are transported in most of the cities of Sub-Saharan Africa. These are run mainly on petrol or diesel, both fossil fuels. Thus, the energy sector is responsible for about 75% of CO_2 emissions, 20% of the methane (CH_4) emissions and a large quantity of nitrous oxide (as in the inorganic fertilizer production industry). Also, the energy sector has been shown to produce nitrogen oxides (NO_x) and carbon monoxide, which are not greenhouse gases, but do have an influence on the chemical cycles in the atmosphere that produce or destroy greenhouse gases (Wikipedia, 2009).

Cement manufacture

The cement production process contributes CO_2 gas. When calcium carbonate ($CaCO_3$) is heated, calcium oxide (CaO) and CO_2 gas are by-products.

Land use

Land use is the particular use to which a given land is put. It could consist of: (1) a major kind of land use, which is defined as a major division of a rural land use (such as rain-fed agriculture, irrigated agriculture, grassland for grazing by ruminant), forestry or recreation; or (2) land utilization type defined as a kind of land use described in degree of detail greater than that of a major kind of land use (example rain-fed annual cropping based on for example groundnut with subsistence maize by small holders; commercial crop production; extensive cattle ranching; softwood plantation by a government Department of Forestry, national park for recreation or tourism, etc.) (FAO, 1976). Other land use types include land where residential, school, church, market, office, industrial, and hospital buildings are erected; other uses include land for road construction, aesthetics and reserved natural forest.

Removal of soil cover (vegetation) would increase the soil surface reflective properties. With the high sand content of soils of the region (Southeastern Nigeria) there would be enhancement of the soil surface albedo. Sand has been reported to have 20–30% albedo (Cunningham *et al.*, 2005). This indicates that the reflected energy from the surface into the atmosphere would contribute to increased atmospheric temperature so that the atmospheric air would become much warmer than it would otherwise be (global warming). Again, if the soil surface is dark following the presence of OM, the implication is that the surface would absorb energy as the surface has low albedo. Absorbed energy heats the absorbing surface to evaporate water. Following the second law of thermodynamics, absorbed energy is gradually remitted as lower quality heat energy. Energy re-released from the earth's warmed surface (terrestrial energy) is lower-intensity, longer-wavelength energy in the far-infrared part of the spectrum. Atmospheric gases, especially CO_2 and water vapor, absorb much of this long-wavelength energy, re-releasing it in the lower atmosphere and letting it leak out to space only slowly. This terrestrial energy provides most of the heat in the lower atmosphere (Cunningham *et al.*, 2005).

Under irrigation agriculture land use, the water applied perhaps through the sprinkler irrigation system tends to increase the relative humidity of the air. The air humidity could be likened to water vapor in the atmosphere. Irrigation operations of any type take place when the air and soil temperature is very high, especially during dry season. Consequently, the applied water would be susceptible to high evaporation, which in turn would go into the air in form of water vapor (also regarded as a greenhouse gas), (Cunningham, *et al.*, 2005; Green, 2008).

About 80% of the earth's natural greenhouse effect has been shown to be due to water vapor and clouds (Green, 2008). Under the flooding type of irrigation, which is a condition that favors rice production or that is witnessed under rice or paddy fields, the soil is covered with water to become anaerobic. Under such conditions, methane – producing bacteria and other organisms – decompose OM in the soil to produce methane (a greenhouse gas), which is emitted into the air to contribute to global warming Brady and Weil, 2005; Wikipedia, 2009).

Manufacture of synthetic compounds with greenhouse effects

Humans release a variety of chlorine – containing molecules into the atmosphere. These are the CFCs and halon gases, commonly known by the trade name – Freon. CFCs were regarded as wonderful compounds. They are nontoxic, non-inflammable, chemically inert and useful in a wide variety of applications. CFCs are used in the developing world as refrigerants, solvents, spray propellants and foam-blowing agents. Until 1978, aerosol spray cans used more CFCs than other products. The molecules are very stable, explaining why they persist for decades or centuries once released into the atmosphere. They are powerful infrared absorbers (Cunningham *et al.*, 2005).

When the CFCs are released into the atmosphere and diffuse out into the stratosphere, the intense UV radiation releases chlorine atoms that destroy the ozone layer. Since the chlorine atoms are not themselves consumed in these reactions, they continue to destroy the ozone layer for many years until they finally precipitate or are washed out of the air. However, the discovery of stratospheric ozone losses has brought about a remarkably quick international response. At a 1987 conference in Montreal, 81 nations agreed to phase out all CFCs (halons, carbon tetrachloride and methyl chloroform). Nevertheless, alternatives to CFCs such as hydro chlorofluorocarbons (HCFCs), (Cunningham *et al.*, 2005), hydro fluorocarbons (HFCs) and per fluorocarbons are already in existence.

Consequences of climate change on Sub-Saharan Africa

Climate changes in many parts of Sub-Saharan Africa, especially Nigeria, are glaring. Globally, increases in temperature following increases in the quantities of the heat – trapping greenhouse gases in the atmosphere (resulting in the earth becoming much warmer than it would otherwise be; that is global warming) has been used as an index of the "Climate change phenomenon". In other words, the changes in the earth's climate driven by the heat from the sun could be detected via changes in temperature (Green, 2008). In the past, climate changes were inferred from changes in indicators that reflect climate such as vegetation, ice cores, dendrochronology, sea level changes, etc.

(Wikipedia, 2009). However, most scientists believe that the average global temperature has increased by 0.5 to 1.0°C during the past century and predict that it is likely to increase by 1.0° to 2.0°C in this century (Brady and Weil, 2005; Cunningham *et al.*, 2005). With these temperature increases, which could be naturally – or anthropogenically – induced, major changes in the earth's climate could result (Brady and Weil, 2005). These changes, which could also be used as indicators of climate change include: changes in rainfall distribution and growing season length, changes in sea levels, heat waves, and frequency and severity of storms, etc. These changes could also act as progenitors of some consequences of "climate change phenomena" on the human societies (people) and natural ecological systems of Sub-Saharan Africa, especially Nigeria. For instance, the anthropogenic changes to the structure, composition or function of large natural bio-geophysical and ecological systems entail changes in the complex array of forcings and feedbacks that characterize the internal dynamics of the earth system (Steflen *et al.*, 2004). These planetary-scale environmental changes have been shown to be of great significance, because they diminish the capacity of the earth's natural environment to supply and replenish resources, and to absorb and recycle the waste products from man's activities. These perturbations are aggravated by global climate change and together they compromise livelihoods by constraining resources in poorer communities and by affecting some goods and services that ecosystems provide (MEA, 2005; Scholes *et al.*, 2007). Africa in general and Sub-Saharan Africa in particular have been reported to be vulnerable to climate change following its location, dependence on natural resources, high levels of poverty, weak infrastructure and limited adaptive capacity (Scholes *et al.*, 2007; Poggos, 2008). However, some observed consequences of climate change in Sub-Saharan Africa especially Nigeria, which could either be positive or negative are as follows:

Precipitation: Rainfall

Reports show that an increase in air temperature makes it possible for the air to carry more water vapor, thus, increased precipitation in the form of rain, snow or hail is theoretically possible (Green, 2008). However, in practice, this relationship of temperature with precipitation has been shown to be more complicated. For instance, some of the humid places such as North Africa are also deserts. But, without the mechanism to condense the water vapor in the air, precipitation would not occur (Green, 2008). A strong month-to-month fluctuations in air temperatures have been observed that occur in the tropics and found that rainfall increased with temperature as is expected in the theory of global warming. In this case, global warming was reported to result in increased water vapor in the atmosphere with corresponding increases in precipitation. However, Green (2008) indicates that there are still significant gaps in the scientific understanding of precipitation systems and their interactions with the climate due to their uncertainty; albeit, at least 80 percent of the earth's natural greenhouse effect is due to water vapor and clouds, which are largely under the control of precipitation. Despite the complications in the relationship between temperature and precipitation (rainfall) and the fact that rainfall is associated with the greatest uncertainty in climate change projections, high influence of climate change in Sub-Saharan Africa could still be observed.

Temporal rainfall variability

It is worthy of note that the rainfall values obtained for a given month of the year and over the entire 30-year period were pooled. It was observed that the mean rainfall values (mm) increased monthly to the month of July (390.21mm) beyond which it slightly decreased in August (346.72mm) (perhaps due to short dry spell characteristic of this month) and then slightly increased in September (393.99mm) and finally decreasing from October (252.23) to December (10.6mm). The rainfall pattern followed the bimodal type of distribution, characteristic of southern Nigeria, especially southeastern Nigeria. This gives rise to the two distinct growing seasons in the region. However, the lower rainfall amount observed in June than in August is unusual as the month of August is characterized by rainfall values lower than those of June, July and September in the region. This deviation from the normal or expected trend could be due to climate change phenomena. It would be recalled that the month of August is characterized by lower rainfall amounts than those recorded for June, July and September in past years due to the short dry spells, popularly known as "August Break" (that could last for about two weeks). However, it was observed that the month of August of 2009 was marked by frequent rainfall events of high amounts and intensity with the characteristic "August Break" becoming extinct and not being observed. This could be attributed firstly to the higher cosmic rays reaching the earth's atmosphere (Green, 2008) during the August month of 2009. Cosmic rays provide material for seeding clouds that make water to condense around the tiny particles, thus, forming more clouds (Green, 2008). Appreciable rainfall amounts have been reported to have been obtained in the tropics from cumulonimbus clouds (Ayoade, 1993). Secondly, much water vapor could have been released terrestrially, that is from the earth's warmed surface into the atmosphere where condensation would eventually take place. Thirdly, much water vapor could have been released in the month of August from the Atlantic Ocean and be taken into the region by the southwest trade winds. In general, the unobserved or unnoticed "August Break" in 2009 is a clear evidence of climate change phenomena.

The high rainfall intensity and amount prevalent in the region (Madubuike, 1999) had at times given rise to flooding incidences that destroyed life and property, especially when they occurred in areas where extreme flooding had not taken place for many years. The Ndiegoro flood disaster in Aba, South-eastern Nigeria, which occurred in the 1980s, is one of such cases. Villages in this area had to relocate following the submergence of their homesteads by water.

It is noteworthy that the ongoing climatic changes arguably are easily linked to the acceleration of the southward encroachment of the Sahara desert, which has engulfed most parts of northern Nigeria like Sokoto, Kano and Borno states.

This is the well-known desert encroachment or desertification (intensifications or extension of desert condition) phenomenon, which has affected one-third of Nigeria's land mass, especially in the northern fringe states (Babalola, 2001). Up till the early 1980s, desertification affected areas northwards of latitude 12^0N. Today, the southward boundary of the desert-like conditions has now moved down to latitude 11^0N as desert encroachment is said to progress southwards at an alarming rate of 1km per annum (Babalola, 2001). Also, Babalola (2001) indicated that one aspect of desertification that requires serious attention is the potential danger, which sand dunes and migrating sands contribute to the rapid spread of the phenomenon. Extensive areas of active dunes and migrating sands exist in many places between Sokoto on the western part and Borno in the eastern part of northern Nigeria. Villages in these areas often have to relocate because of the blasts of migrating sands on their homesteads; high ways are regularly blocked and establishment of crop and tree seedlings have been hampered because of poor establishments as they often get buried by sand. Reports show that the land area under sand dunes has increased over the years and is still increasing at a rate of 200 km^2 per annum within the areas affected by desertification (Babalola, 1999 and 2001).

Floods, drought, desertification and other weather extremes have adverse effects on agriculture (through loss of soil quality and domestic animals), which would in turn give rise to food insecurity, poverty, social tensions and migration of people (to urban areas or other countries) (Poggo, 2008) and also break-down in human health; loss of water resources and natural resource-based livelihoods (Scholes *et al.*, 2007), especially in Sub-Saharan Africa.

Effects of climate change on agriculture

Climate has enormous effects on vegetation and animal life even though man's activities can affect climate by introducing pollutants into the environment. Agricultural trade between regions has also provided substantial amount of food to import nations. At the same time, it has equally provided comfortable income to the exporters of agriculture products. It is thus expected that climate change will affect the productivity of agricultural products in terms of quantity and quality of yield.

The increased rainfall incidence following climate change phenomena would result in high soil erosion processes by water, which would in turn give rise to on-site impacts, such as: soil and plant nutrients losses (Foster *et al.*, 1985; Lal, 1976 and 1980), loss of plant available soil water (Masses and Waggoner, 1985), desiccation and loss of arable land, especially through gully erosion (Aneke, 1985), and then off-site impacts such as sediments transport and deposition into streams and rivers, causing siltation and pollution of water bodies (Klik, 2000), water quality decline (Richards, 2002) and eutrophication (Gruhn *et al.*, 2000) affecting aquatic habitats, including fish (Brady and Weil, 2005) adversely. These would result in reduced food production, which would in turn lead to hunger, malnutrition, poverty, social tensions and migration, and health breakdown and vices.

Climate change phenomena could have some beneficial effects on agriculture. For instance, increases in temperature will quicken the development of crops thereby reducing the duration between sowing and harvesting in the case of annual crops. However, the shortening of the normal cycle could have undesirable effects on productivity as senescence would occur faster.

Carbon dioxide (CO_2) is important for plant growth. CO_2 increases in the atmosphere could have positive and negative impacts. For instance, there could be positive physiological effects by an increase in photosynthesis. Thus, CO_2 is a fertilizer for plants and higher CO_2 levels bring about faster growth for many plants species (Cunningham *et al.*, 2005). Conversely, higher CO_2 levels would lead to decreased nitrogen uptake by plants. This would result in crops with reduced nutritional values. Decreased amount of nitrogen in grazing plants has been shown to reduce productivity in sheep.

Quality of products could also be affected by climate change. In February 2001, the Intergovernmental Panel on Climate Change (IPCC) released its Third Assessment Report, which indicated with a high degree of confidence that recent changes in the world's climate have had discernible impacts on the physical and biological systems (IPCC, 2001). According to the IPCC's Third Assessment Report for rice, the amylase content is increased under increased CO_2. However, the zinc and iron concentrations, which are important for human nutrition, would drop lower (Seweera and Conroy, 1997). The protein content of this grain has also been reported to decrease under combined increases of temperature and CO_2.

Increases in the level of some streams and rivers, often witnessed during the rainy season, which at times could be in excess and would cause inundation of low lying lands, could result in loss of agricultural lands. This situation is common in southeastern Nigeria.

Human health problems

Global warming could increase rainfall in certain areas giving rise to an increase in the atmospheric humidity and wet season periods. In combination with high temperature, this condition could favor the development of malaria, bacteria, and fungal diseases. Also, there could be an increased pressure from insects and disease vectors as a result of higher temperatures and humidity.

Malaria incidence, meningitis and other skin debilitations now appear in places where they have never existed before (Glantz, 2005; Green 2008). In southeastern Nigeria, malaria incidence is often high during the rainy or wet seasons due to presence of conditions that tend to favor the breeding of mosquitoes (Vector of malaria). Climate change may alter the distribution of important vector species and this may increase the risk of introducing diseases into new areas. Altered wind patterns may change the spread of both wind-born pests and diseases, and of bacteria and fungi that are agents of some diseases. Another human health problem associated with climate change is heat stress (due to exposure to high temperature). Stressful hot weather episodes (heat waves) tend to cause deaths in the elderly. It is noteworthy that global mental-health problems have also been associated with climate change. Stratospheric ozone losses have been associated with increased CFC gases (greenhouse gas of industrial origin). If this situation is allowed to continue, there would be numerous health problems. The ozone layer absorbs some wavelengths of ultraviolet solar radiations known as UV-B (290–330). This absorbed energy makes the atmosphere warmer towards the top of the stratosphere. Though

UV radiation damages living tissues, its absorption in the stratosphere also protects life on the surface of the earth. So ozone depletion implies increases in unabsorbed UV radiation that could cause high rates of skin cancer, genetic mutation, crop failures and disruption of important biological communities (Cunningham *et al.*, 2005). This has therefore called for urgent adoption of some strategies to mitigate global warming and enhance the adaptive capacity of people of Sub-Saharan Africa, to global warming.

Mitigation and adaptation strategies

Climate change is a global environmental problem, thus, calling for global solutions. This would explain why several world summits have been taking place since 1992 to address the problem. Recently, another conference that ended on the 18th of December, 2009, was held in Copenhagen, Denmark, to find lasting solutions to global warming but reached a deadlock. However, doing nothing towards this is not an option. This is therefore the time for Africans, especially Sub-Saharan Africans to rise and face up to global warming.

However, to effectively tackle this problem, efforts should first be geared towards reducing or eradicating the causative agents of global warming, that is the greenhouse gases. Some of the mitigating strategies could be in form of replacing use of fossil fuels with renewable energy sources such as solar power, wind power, geothermal energy, water power, biomass and fuel cells. Nuclear power is also being promoted as an alternative to fossil fuels. It is true, nuclear reactions do not produce greenhouse gases, but security worries and unresolved problems of how to store wastes safely make this option unacceptable to many people (Cunningham *et al.*, 2005). The use of wood-based products are also alternative sources of power and could be used in combating climate change; as wood is man's only naturally renewable raw material. End-of-life wood products can be reused, recycled or valued as energy sources. Combustion of wood residues is an O_2-neutratralization operation, hence additionally mitigating climate change by reducing fossil fuel consumption. In addition to reducing its output; there are options for capturing or sequestering CO_2 or carbon. Planting trees can be effective if they are allowed to mature. Trees like *Gmelina arborea, Azidirachta indica, Delonixregia, Gliricidia sepium, Leucaene leucocephala,* etc., could be used. Farmlands can also serve as carbon sinks if farmers change their crop mixtures and practice minimum tillage that keeps carbon in the soil. It is worthy of note that most attention is focused on CO_2, because it lasts in the atmosphere on average for about 120 years. Methane and other greenhouse gases are much more powerful heat absorbers but remain in the air for a much shorter time (Cunningham *et al.*, 2005). Reducing gas pipeline leaks would conserve these valuable resources as well as help the environment. Methane from landfills, oil wells and coal mines that once would have been vented into the air is now being collected and used to generate electricity. Rice paddies are a major methane source. Changing flooding schedules and fertilization techniques can reduce anoxia that produces marsh gas. Ruminant animals (like cows, camels and buffalos) create large amounts of digestive system gases. Modified diets can reduce flatulence significantly. Reducing soot from diesel engines coal-fired generators, forest fires and wood stoves might reduce global warming. There should also be government regulations in such areas as construction and transport in order to improve energy efficiency. The developed countries should provide support to Sub-Saharan Africans considering the fact that they have low adaptive capacity and that the region is vulnerable to climate change.

Conclusion

Irrespective of the actions taken to mitigate greenhouse gas emissions or otherwise address the causes and consequences of global climate change, one of the most important responses to global climate change will be to deliver timely information and current useful and relevant scientific findings to decision-makers and the public. The best scientific information on the patterns, projections and consequences of both climate variability and change should be widely used, together with an understanding of associated uncertainties and should lead to international policy making that protects the environment, enhances socio economic development and ensures a sustainable future for the nations.

The governments and general public, therefore, need to understand that the world inhabited by humans is variable and prone to change, thus reflecting the dynamic nature of the world. Also, they need to become aware of what steps and actions each nation could undertake to reduce human impact on the global environment.

Table 1: Descriptive Statistics: [mean (mm), Standard deviation (mm) and coefficient of variability CV (%) values] obtained from pooled rainfall amount (mm) data of the 12 months (January – December) under each year of the 30 – year (1980–2009) period and then each month of the entire 30 -year period. Overall means CV% = 85.21.

Year	Number (N)	Minimum (mm)	Maximum (mm)	Mean (mm)	Standard Deviation (mm)	CV (%)	Months	Number (N)	Minimum (mm)	Maximum (mm)
1980	12	0.0	497.10	195.83	160.91	82.34	January	30	0.00	105.3
1981	12	0.0	543.9	197.83	168.96	85.41	February	30	0.00	110.6
1982	12	0.0	429.9	185.35	147.98	79.84	March	30	7.1	259.9
1983	12	0.0	573.5	146.22	173.83	118.88	April	30	36.0	354.2
1984	12	0.0	372.8	179.69	142.46	79.28	May	30	47.30	542.5
1985	12	3.6	453.2	202.78	159.16	78.49	June	30	0.0	513.1
1986	12	0.50	546.0	206.89	171.63	82.96	July	30	223.0	650.0
1987	12	0.0	475.0	173.0	168.51	97.40	August	30	157.9	641.0
1988	12	10.90	446.1	213.64	162.84	76.22	September	30	193.9	690.2
1989	12	0.0	469.5	215.13	175.37	81.52	October	30	23.30	399.20
1990	12	7.10	650.10	227.66	227.56	99.96	November	30	1.50	167.8
1991	12	0.0	521.10	213.24	190.74	89.45	December	30	0.00	52.4
1992	12	0.9	489.9	197.84	155.00	78.35	–	–	–	–
1993	12	0.0	464.3	181.66	140.17	77.16	–			–
1994	12	0.0	690.2	234.05	230.06	98.29	–	–	–	
1995	12	12.50	484.0	218.52	184.35	84.36	–	–	–	–

Year	Number (N)	Minimum (mm)	Maximum (mm)	Mean (mm)	Standard Deviation (mm)	CV (%)	Months	Number (N)	Minimum (mm)	Maximum (mm)
1996	12	0.0	573.2	224.97	200.41	89.08	–	–	–	–
1997	12	0.0	542.7	240.83	174.64	72.51	–	–		–
1998	12	0.0	290.3	136.79	117.84	86.14	–	–		–
1999	12	0.0	609.0	209.62	166.73	79.73	–	–	–	–
2000	12	0.0	391.8	194.77	164.19	84.30	–	–	–	–
2001	12	5.50	430.8	192.03	149.69	77.95	–	–	–	–
2002	12	0.0	391.5	156.87	160.87	102.56	–	–	–	–
2003	12	0.20	476.4	197.53	165.80	83.94		–	–	–
2004	12	0.0	334.6	159.71	125.79	78.76	–	–	–	–
2005	12	11.80	334.6	226.06	185.79	82.19	–	–	–	–
2006	12	0.0	513.1	234.23	192.75	82.29		–	–	–
2007	12	0.0	432.0	188.13	144.61	76.86	–	–	–	–
2008	12	0.0	524.5	209.48	170.86	81.56	–	–	–	–
2009	12	0.0	542.5	235.96	208.86	88.52		–	–	–

References

Al-Hamndou, D., 2008. Climate change worsens weather extremes. In: Palus, N., VOA News Reporter: Migration and development in Africa. Conference held in Dakar, Senegal. March 20, 2008. VOANews@ OANews.com.

Ayoade, J. O., 1993. Introduction to climatology for the tropics. Spectrum Books Limited, Ibadan. Pp. 1–247.

Babalola, O., 1999. Development of technology for sand dune stabilization in Northern Nigeria. In: Dov, P. and A. Schlissel (eds.). Combating desertification with plants. Kluwer Academic/Plenum Publishers. New York. Pp. 347–358.

Babalola, O., 2001. Nigerian Agriculture: Basis for hope, hurdles against hope, hope for tomorrow. 2001/2002 University Lectures, University of Ibadan. Pp. 33–52.

Benson, T., 2004. Africa's food and nutritional security situation. Where we are and how did we get here? 2020 Discussion Paper 37. International Food Policy Research Institute, Washington D. C.

Boesch, D. F., Josselyn, M. N., Mehta, A. J., Morris, J. T., Nuttle, W. K., Simenstad, C. A. and D. J. P. Swift, 1994. Scientific assessment of coastal wetland loss, restoration and management in Louisiana. Journal of Coastal Research Special Issue, 20: 1–89.

Boko, M., Niang, L., Nyong, A., Vogel, C., Githeko, A., Medani, M. and B. Osman-Elash, 2007. Africa Climate Change, 2007, "Impact, Adaptation and Vulnerability". In: Parry, M. L., Canziani, O. F., Palutikof, J. P., Van der Linden, P. J. and C. E. Hanson (eds.). Fourth Assessment Report of the Intergovernmental Panel on Climate Change. Cambridge University Press, Cambridge.

Brady, N. C. and R. R.Weil, 2005. The nature and properties of soils 13th edition. Pearson Education (Singapore). Pte. Ltd. Indian Branch, Patpargang Delhi, India. Pp. 537–553.

Charlson, R. J., Schwartz, S. E., Hales, J. M., Cess, R. D., Coakley, J. A., Hansen, J. E. and D. J. Hofman, 1992. Climate Forcing by Anthropogenic Aerosols. Science: 255: 423–430.

Crooks, S. A., 2005. Characterization of the 11-year solar signal using a Multiple Regression Analysis of the ERA-40 Data-Dataset. Journal of Climate. 18: 821–996. doi: 10. 1175/JCL- 3308.1.edit.

Cunningham, W. P., Cunningham, M. A. and B.W. Saigo, 2005. Environmental Science: A global concern. 8th edition Mcgraw-Hill Higher Education Companies, New York. Pp. 306–352.

Federal Department of Agriculture and Land Resources (FDALR), 1985. The reconnaissance and soil survey of Rivers State, Nigeria (1: 250,000). FDALR, Soils Report, P. 215.

Fischer, G., Shah, M. and H. van Velthuizen, 2002. Climate change and agriculture vulnerability. Report prepared for the World Summit on Sustainable Development. International Institute for Applied Systems Analysis, Vienna. P. 152.

Fischer, G., Shah, M., Tubiello, N. and H. van Velthuizen, 2005. Socio- economic and climate change impacts on agriculture on integrated assessment, 1990–2008. Philosophical Transactions of the Royal Society, 360: 2067–2083.

Foley, J. A., De Fries, R., Asner, G. P., Barford, C., Bonan, G., Carpenter, S. R., Chapin, F. S., Coe, M. T., Daily, G. C., Gibbs, H. K., Helkowski, J. H., Holloway, T. N., Howard, E. A., Kucharik, C. J., Monfreda, C., Patz., J. A., Prentice, I. C., Ramankulty, N. and P. K. Snyder, 2005. Global consequences of land use. Science, 309: 570–574.

Food and Agriculture Organization (FAO), 1976. A Framework for Land Evaluation. FAO Soils Bulletin, No. 32. FAO of the United Nations, Rome, Italy. Pp. 11–16.

Food and Agricultural Organization (FAO), 1994. Soil map of the world legend: FAO, United Nations. Rome. P. 131.

Foster, G. R., Yaung, R. A., Romkeus, M. J. M. and C. A. Onstad, 1985. Process of soil erosion by water. In: Follet, R. A. and B. A. Stewart (eds.). Soil erosion and crop productivity. Agron Soc. Am, Inc.; Crop Sci. Soc. Am, Inc.; Soil Soc. of Am., Inc., Publishers: Madison, Wisconsin, USA. Pp. 137–162.

Galloway, J. N., Dentener, F. J., Capone, D. G., Boyer, E.W., Howarth, R. W, Seitzinger., S. P., Asner, G. P., Cleveland, C. C., Green, P. A., Holland, E. A; Karl., D. M, Michaels; A. F., Porter, J. H; Townsend, A. R and C. J. Voros Mary, 2004. Nitrogen cycles past, present and future. Biogeochemistry, 70: 153–226.

Glantz, M. H., 2005. Climate change (Global Warming) and Sub-Saharan Africa. Fragile ecologies.

Green, K. C., 2008. Global climate change: Evidence on causes and effects. A report prepared for Morgan Family Charitable Foundation. Pp. 4–97.

Gyampoh, B. A., Amisah, S. and M. Idinoba, 2008. Coping with climate change: how local communities use traditional knowledge in rural Ghana. Paper presented at: Adaptation of forests and forest management to changing climate with emphasis on forest health: A Review of Scene Pokies and Practices, Urnea, Sweden.

Gruhn, P., Golelti, F. and M. Yudelman, 2000. Integrated nutrient management, soil fertility and sustainable agriculture: Current issues and future challenges. Food, agriculture and the environment. Discussion paper No. 32., International Food Policy Research Institute, Washington, D.C., USA.

Intergovernmental Panel on Climate Change (IPCC) (2001). Climate change. IPCC Third Assessment Report. www.grida.no/climate/ipcctar/. Intergovernmental Panel on Climate Change (IPCC), 2007. Climate change. The physical science basis. Summary for Policy Makers (IPCC).

Klik, A., 2000. Soil erosion on agricultural land. In: Stoecker, B. (ed.). Ecosystem service and sustainable watershed management in north China. International Conference. Beijing, P. R. China. August 23–25, 2000. Ankeny. Pp. 163–185.

Lal, R., 1976. Soil erosion problems on an Alfisol in Western Nigeria and their control. IITA Monograph, Ibadan.

Lal, R., 1980. Losses of plant nutrients in runoff and eroded soil. In: Rosevall, T. (ed). Nitrogen Cycling in West African Ecosystems. Pp. 31–38.

Lekwa, G. and F. P. Whiteside, 1986. Coastal plain sands of southeastern Nigeria. Morphology, classification and genetic measurement. Soil Science Society of American Journal, 50: 154–156.

Madubuike, C. N., 1999. Comparative analysis of soil loss in southeastern and southwestern Nigeria. Journal of Erosion and Environmental Degradation. 1(1): 43–52.

Masses, T. W. and H. O.Waggoner, 1985. Productivity losses from soil erosion on dry crop land in the inter mountain area. J. Soil and Water Cons. 40: 447–454.

McCarthy, J. J., Canziani, O., Leary, N. A., Doken, D. J. and K. S. White, (eds.), 2001. Climate change 2001 – impacts, adaptation and vulnerability. Contribution of working Group II to the Third Assessment Report for Intergovernment Panel on Climate Change. Cambridge University Press. Cambridge.

MEA, 2005. Ecosystem and human well-being; Current state and trends. Millennium Ecosystem Assessment. Vol. 1. Island Press, Washington, D. C.

Ofomata, G. E. K., 1975. Soils. In: Ofomata G. E. K. (ed.). Nigeria in Maps: Eastern states. Ethiope Publishing House, Midwest Mass Communication Cooperation, Benin City, Nigeria. Pp. 41–42.

Ofurum, S. C. (ed.), 2009. Climate change. Effect on mental health worries researchers. In: The Leader Newspaper, December 13–20, 2009. Printed and Publisher by Assumpta Press, Owerri, Nigeria. P. 15.

Poggo, A., 2008. Consequences of climate change in Sub-Saharan Africa. A paper presented to the self select group on climate change, Lambeth Conference, July 31, 2008.

Scholes, B., Ajavon, A., Nyong, T., Tabo, R., Vogel, C. and I. Ansorge, 2007. Global environmental change (including climate change and adaptation), In: Sub-Saharan Africa. Prepared for International Council for Science, (ICSU). Regional office for Africa. Pretoria, Republic of South Africa.

Siegenthaler, U., Stocker, T. F., Monnin, E., Luthi, D., Barnola, J. M., Fischer, H., Masson-Delmotte, V. and J. Jouszel, 2005. Stable carbon cycle – climate relationship during the late Pleistocene. Science. 310: 1313–1317.

Spahni, R., Chappellaz, J., Stocker, T. E., Loulergue, L., Hausammann, G., Kawamura, K., Fluckiger, J., Schwander, J., Raynaud, D., Masson- Delmotte, V. and J. Jouzel, 2005. Atmosphere methane and nitrous oxide of the late pleistocene from antarctica ice cores, Science, 310: 1317–1312.

Steflen, W., Sanderson, A., Tyson, P., Jager, J., Matson, P., Moore, I. B., Oldfield, F., Richardson, K., Schellhuber, H. J., Turner, I. I. and B. L. Wasson, 2004. Global change and the earth system: A planet under pressure. Springer-Verlag, Berlin and Heidelberg.

Stern, N., 2007. The economics of climate change. The Stern Review. Cambridge University Press. Cambridge.

Wikipedia, the free encyclopedia, 2009. Climate change.

Yale Environment 360, 2009. Mega droughts in Sub-Saharan Africa. A publication of Yale School of Forestry and Environmental Studies.

Gabriel O. Ihejirika
Federal University of Technology, Owerri, Nigeria

Awareness of the causes and environmental implications of climate change phenomena in select-countries of Sub-Saharan Africa

Abstract
Adaptation, mitigation and geo-engineering are strategies to climate change and variability. The earth's climate is dynamic and always changing through a natural cycle. What the world is more worried about is that changes that are occurring today have been speeded up because of man's activities. These changes are being studied by scientists who are finding evidence from fossil fuels: sources such as oil drilling, coal mining and natural gas supply, and energy sectors, which are responsible for about 75% of CO_2 emissions, 20% methane emissions, and nitrous oxide, and carbon monoxide. These gases have great influence in the atmosphere by producing or destroying greenhouse gases causing global warming. Specifically, a number of requirements must be attained for enhancing adaptive capacity. These include: improved access to resources, reduction in poverty, lowering of inequities in resources and wealth among groups, improved education and information, diminished inter-generational inequities, respect for accumulated local experience, moderate long-standing structural inequities, assurance that responses are comprehensive and integrative, not just technical, and active participation by concerned parties, especially to ensure that actions match local needs and resources. While ensuring that any use of natural resources is equitably and ecologically sustainable, the integrity and diversity of nature must be protected. Providing a "tool kit" from which to diminish the vulnerability of human systems to climate-related hazards in Sub-Saharan Africa becomes necessary. Both traditional communities and modern nation-states have developed disaster risk mitigation tools based on the conservation of natural systems that provide "protective services."

Introduction

The impact of our changing climate on human systems is the rise in incidence and severity of climate-related disasters. After decades of skepticism about global warming and its after effects, the world has come face-to-face with the realities of that danger particularly in fragile tropical landscapes where the majority of the world's poorest people live on peasant agriculture. The United Nations (UN) in 2008 during her General Assembly Proceedings warned that global warming could cost the world up to $20 trillion over the next 20 to 25 years. This warning was made in order to place the world on a markedly different and sustainable energy trajectory. The report also warned that the change could do the most harm to people who least could adapt to global warming. The challenge of climate change is unlikely to be gender neutral, as it increases the risk to the most vulnerable and less empowered social groups. The General Assembly debate was intended to shape overall UN policy on climate change, including how nations could adapt to a warmer world and find ways of supporting the UN-led negotiations that were expected to have been agreed by 2009. The treaty replacing the Kyoto Protocol that expires in 2012 was expected to shape the course of climate change for decades to come. The climate is already changing and it represents one of the greatest environmental, social and economic threats facing Sub-Saharan Africa. There is an international scientific consensus that increasing levels of man-made greenhouse gases are leading to global climate change. People are now aware that the causes of climate change can be divided into two categories: those due to natural and those due to man-made causes.

The man-made causes include greenhouse gases and their sources. CO_2 is undoubtedly the most important greenhouse gas in the atmosphere. Changes in the use patterns, deforestation, land clearing, agriculture, and other activities have all led to a rise in the emission of CO_2. Electricity is the main source of power in the urban area. All our gadgets run on electricity generated mainly from thermal fuel plants. These power plants are mainly dependent on fossil fuels (mostly coal) and are responsible for the emission of large amounts of greenhouse gasses and other pollutants.

Cars, buses and trucks are the principal ways by which goods and people are moved in most of our cities. These are run mainly on petrol or diesel, both are fossil fuels. We generate large quantities of wastes in the form of plastics that remain in the environment for many years and cause one damage or the other. Large areas of forests are cut down as a large quantity of timber is used in the construction of buildings and houses. There is ecological degradation as the land shrinks, because the area of land used for agriculture is limited.

High yielding varieties of crops are grown to increase the agricultural production. These require large quantities of fertilizers and organic manure, control of weeds, pests and diseases using chemicals with their associated problems as they mean more emissions of nitrous oxide, both from yields in which it is put and chemical industries that produce them. Pollution results from the run off of fertilizers, herbicides, pesticides, and fungicides into water bodies.

Cultivated land areas are diminishing and environmental and biological resources are being degraded and destroyed. Sub-Saharan Africa is faced with economic threats to its food security, because the climate change and its causes will reduce food surpluses in the region, raise grain prices, shrink food aid with its cutbacks in foreign assistance as those countries that would have assisted with foreign aid would also be seriously be affected by climate change.

Worst still is the increasing proportion of allocations to disaster assistance of certain areas thereby reducing the funds available for agricultural research investments of some other areas. These factors suggest that developing countries will face growing food deficits, food insecurity and nutritional insecurity. They may also face environmental degradation and natural resources scarcities that will end up in greater competition and conflicts.

In some cases man-made causes of climate change remain, especially where prolonged conflicts prevent farming and marketing and where land, waterworks, markets, infrastructure and human communities have been destroyed. This is evident in form of massive destruction of natural vegetation, soil structures and composition, biotic and abiotic agents that affect the natural ecosystem, causing environmental degradation and global warming.

Also entire regions are destabilized politically and economically by refugees from a series of civil wars by people, who, self-settled or maintained in refugee camps, transform the political economy of the region, added to a burdensome military presence. The flood of indigenous people fleeing military brutalities are a challenge to the economic and political stability of some parts of Sub-Saharan Africa that seek to resettle refugees and maimed people away from volatile political areas. Contributing to meet of food needs of refugees places a particular burden on recipient communities where food security is already marginal. Additional demands to overburdened communities for food, water, land, and fuel can reduce households that were marginally food secure to acute to those with food shortages. In most cases, households that are forced to sell assets to buy food often find markets disrupted, forage areas stripped, and buffers such as livestock reduced in value, because refugees fleeing the fighting, upset markets transforming food shortages into a 'famine that kills'.

There is increased destruction of food stocks, livestock, and other assets in food-producing regions, cutting off market supplies of food in these and other regions, and diverting food relief from intended beneficiaries to the military and their supporters. Farming populations are also reduced by direct attacks, terror, enslavement, or forced recruitment and by malnutrition, illnesses and death. As farming populations flee, decline, or stop farming out of fear, production falls, thus spreading food deficits over wider areas. Land-mining and poisoning of wells are additional hostile acts that turn temporary acute food shortages into long-term insufficiencies. This discourages and forces people to leave and not return and thus interrupt food production and economic activities permanently.

Some of the interventions to these man-made causes and consequences of climate change include: efforts of international communities to maintain geographical information on famine, early warning systems and food reserves, where information systems identify impending or actual acute food shortages; the United Nations and bilateral donors with the International Federation of Red Cross and Red Crescent Societies (IFRC), the World Food Program (WFP), and nongovernmental organizations (NGO), try to come to the aid of those regions seriously affected by climate change.

While international negotiations aimed at reducing the sources of climate change proceed, efforts to respond to the consequences must likewise be developed and implemented. But, what are the best actions to take in responding to the consequences of climate change? Facing the realities of competing priorities and limited resources, actions are more likely to be implemented if they offer benefits that outweigh the costs. Estimating the benefits of particular measures in response to climate change may be difficult or impossible. For that reason, those measures whose non-climate change-related benefits exceed the costs are the measures most likely to be adopted and implemented. These are options offering co-benefits, and their identification and promotion is a primary goal. Natural resource mismanagement contributes to the vulnerability of human systems to disasters, while enhanced management can provide a tool for vulnerability reduction. If so, targeted conservation of particular natural buffer systems will in many circumstances offer no-regrets adaptation opportunities. The tools for implementing effective management of natural areas have been developed over a century of experience by conservationists, as typified by the International Union of Conservation of Nature (IUCN).

Climate change and environmental implications in tropical landscapes

With reports of extreme floods in Kenya, Nigeria, Indonesia, Thailand, Southern Africa and India, the world has been jolted with the stark reality that it is heading towards something worse if urgent remedial steps are not taken. Also, manifested are the increased cases of drought in the Sahelian region of Africa and the now common incidence of hurricanes in the United States of America.

In central Nigeria, excessive warming is being blamed for the consistent violence between nomadic cattle herders and farmers. The duo has been locked in battle for the possession of the scarce green areas as the desert creeps southwards. Similarly, deforestation, dwindling water supplies and rising sea levels are already sparking mass migrations, and provoking ethnic conflicts. UN reports on climate change indicate that humanity may have to brace for more challenges as they predict a rise in temperature in the order of 1.4 degrees to 5.8 degrees Celsius in Africa.

Regions that are already least secure in food production, like Sub-Saharan Africa stand to be the worst victims of global warming as wet areas become wetter and dry areas become drier. Speaking on the issue, Ms Jennifer Morgan, the Director of Global Climate Change Program at the WWF, in a review of a recent international report on the work of 1.360 scientists in 95 nations, described Africa as the most vulnerable continent that will find it increasingly difficult to cope as desertification is threatening to drive the populations from their regions. She particularly cited an instance in Uganda where researchers and the government found that the climate has become hotter. It was also found that the rains were becoming even more erratic in the last decade, posing a threat to the key coffee crop.

To justify their fears of the worst for Africa, Abutu (2008) mentioned that environmentalists point to gullies of eroded barren earth scarring the shorelines of Lake Victoria, which borders Kenya, Uganda and Tanzania. Rising sea temperatures are also among the threats to Africa's lush east coast, the main survival line of poor coastal communities who depend solely on fisheries and tourism. Morgan (1986) sums up the effects as follows: "On the whole, the results have been loss of lives, destruction of property, injury and hardship inflicted on humidity, underscoring the fact that global warming is both a reality and a phenomenon that begs for collective action."

Lake Chad, a major water reservoir for the West and the Central African region, has shrunk substantially within the last two decades with great attendant socio-economic and environmental implications for the surrounding countries of Nigeria, Chad, Niger, Central African Republic, and Cameroon.

Nutritional insecurity

Individual food deprivation, or nutritional insecurity, a protein-energy or micronutrient mal-nutrition, may affect individuals even in situations where a band of community households appears to be food secure. Women, children, the elderly, or socioeconomically marginal members of households such as servants or those with no ethnic minority identity, may be deprived of adequate food or suffer malnutrition even where household food supplies are adequate or plentiful. In situations of active conflict, women and children or those who are left behind may have less access to food after men are mobilized into armed forces or migrate in search of additional food or employment. They also face high risks of illness and malnutrition when health care services and social service institutions are destroyed. In the event of emergency, food rations may be nutritionally unbalanced and insufficient to meet demands in the absence of additional markets and sources of food and income to maintain their food supplies, health, and care are jeopardized.

Response to climate change

There has been continuous political and public debate regarding climate change, and what actions to take in response to it. In the face of climate change, the global communities, nations and local communities of Sub-Saharan Africa are undertaking actions along two primary tracks: Mitigation, the process of reducing greenhouse gas emissions and associated climate change, and adaptation, the process of adjusting in response to, or in anticipation of climate change and more specifically, genetic engineering, to reverse global warming. Most national governments have signed and ratified the Kyoto Protocol aimed at reducing greenhouse gas emissions.

Community adaptation to climate change extremes and variability

There is much to be learned from examining traditional means of living with natural hazards. Traditional societies typically lack the means to dramatically modify their environment, and therefore have developed livelihoods and coping strategies adapted to variability. Farmers practice intercropping, homes are built in locations known to be sheltered from storms and avalanches, housing is made to fail without killing its occupants.

Soil management

This approach for increasing the stability and productivity of soil is a general term that involves a range of specific techniques such as fallow cycling, forest buffering, selective planting, managed grazing, etc. Soil management is recognized as central to combating desertification.

Water harvesting

Around the world, this approach has many variants, from the construction of johads (earthen dams) in Northwest India, to rooftop collection in the Caribbean, to the forest Zai in West Africa. Water harvesting techniques have been used as a drought-proofing tool to increase water availability for households, irrigation, as well as baseline water flow for watershed restoration. (In Western Sudan, for instance, the trunk of the *Adansonia digitata* tree is used for water harvesting, purification and storage).

Intercropping

The technique of planting selected food crops within stands of trees (e.g., in the case of Sudan, gum Arabic stands) can provide local communities with added food security and income through livelihood diversification, while at the same time reducing deforestation and desertification. An example is the coastal mangrove and the resulting protection from storm surges in Vietnam and of the agro ecological practices reducing the impacts of landslides and flooding in the wake of Hurricane Mitch in Honduras, Nicaragua and Guatemala.

Climate change and variability are a considerable threat to agricultural communities, particularly in lower latitudes. This threat includes the likely increase of extreme weather conditions, increased water stress and drought, desertification, as well as adverse health effects (extreme heat and increased diseased spread of diarrhea and infectious disease, as well as malaria). Hillside reforestation to mitigate landslides, wetland conservation and restoration to provide storm surge buffers, mangrove protection to avoid coastal flooding and removal of alien species of plant to restore hydrological conditions are some of the measures for combating climate change.

Switzerland learned the importance of conserving its high mountain forests over 100 years ago. Widespread flooding, avalanches and landslides in the late 19th century demonstrated the link between deforestation and these catastrophic events and led to the passage in 1876 of the historic Switzerland law aimed at conserving mountainous forested areas. The protective value of these forests include: protecting villages, towns, infrastructure and tourism, and thus the economy as a whole. Likewise, in Bangladesh it is observed that the Sundarbans, one of the world's largest mangrove forests, protect several communities from the otherwise-devastating tropical cyclones. An estimate by the Bangladeshi government suggests that the storm protection provided by the mangrove area has saved the public purse nearly half a billion dollars in avoided costs of structural destruction and relocation. In Central America, the seven disaster-prone countries of the region are collaborating to protect a Mesoamerican Biological Corridor along the Atlantic coast of the isthmus, from Mexico to Panama. The corridor aims to preserve watersheds and forests, while attracting over US$100 million per year in international greenhouse gas offset funds.

Natural resource management and climate change adaption

The application of targeted conservation of natural buffer systems as a strategy for adapting to climate change offers several potential co-benefits: biodiversity conservation, poverty alleviation and enhanced CO_2 sink capacity. Biodiversity provides many direct benefits, including the natural resource base essential for many livelihoods, genetic material for breeding and new medicines, aesthetic beauty, and tourism revenues. Moreover, according to the World Conservation Monitoring Centre (1992), the diversity of species under girds the ability of an ecosystem to provide most of its other goods and services. Reducing the biodiversity of an ecosystem may well diminish its resilience to disturbance, increase its susceptibility to disease outbreaks and decrease productivity. Thus, in addition to directly providing livelihoods, biodiversity is important in ensuring the long-term viability of natural buffer systems in the face of disasters.

Benefits to the poor and most vulnerable

Adaptation measures focused on enhanced natural resource management can also benefit the poor more effectively than large-scale structural measures. It is the poor who are most dependent on natural resources, both directly for their livelihoods, and for security during times of crisis. It is also the poor who are most often living on marginal lands and in vulnerable locations, who would benefit most directly from enhanced resource management. The World Bank (1988) estimated that poverty increases in proportion to distance from cities and coasts, typified by the distribution of poverty in China, Vietnam and Latin America. Agriculture, forestry and fishing are directly responsible for 50% of all jobs worldwide and 70% of the jobs in Sub-Saharan Africa, East Asia and the Pacific. Because the poor have a narrow and geographically concentrated set of livelihood sources, they are most vulnerable to losing their livelihoods during disasters.

Coping strategies typically employed in the wake of a disaster include changing the mix of livelihoods, creating new livelihoods, seeking new sources of resources (by force of arms or otherwise), and mitigation. A group or individuals will seek to deploy the different assets they possess to best effect within the range of choices they possess. The poor usually have the least choice among strategies, receive the least assistance from government authorities and are therefore most dependent on the state of the environment for providing alternative livelihoods. Thus, investing in the natural resource base that sustains their livelihoods may have a direct positive impact on their immediate lives and long-term resilience to climate variability.

Enhanced CO_2 sink capacity

A number of environmental management-based adaptation activities can also serve as climate change mitigation measures. For example, sustainable rangelands management activities have in the past been employed in a number of countries to combat desertification, a local, national and regional sustainable development goal. Recently, these measures have been explored in trees for their simultaneous drought-proofing attributes to channel climate adaptation and carbon sequestration capabilities – central endeavor of climate change mitigation.

Climate change adaptation

After years of being relegated to the peripheries of climate change discussions, adaptation issues were given greater attention through a number of key developments (under UNDP/GEF) of new adaptation assessment guidance known as the Adaptation Policy Framework (APE) and the Marrakech Accord, which created three new funds to assist developing countries to adapt to climate change (UNDP, 2002 as well as RIOA, 2008; Noss, 2001; FOA, 2007).

Slash and burn agricultural practice

It is against this backdrop that stakeholders continue to design strategies to contend with the challenges of how to mitigate the effects of global warming.

Expectedly, more attention has been on such vulnerable areas as Africa, already scarred by deforestation, poor soil health and drought. The strategies are contained in some protocols and conventions, among which are the Montreal and Kyoto Protocols, targeted at reducing the emission of carbon dioxide globally. The campaign, fashioned after the works of a Nobel Laureate, Prof. Wangari Maathai, and sponsored by the UN Environment Program (UNEP), hopes to realize the target of planting one billion trees by the end of 2007. Maathai said that the vital importance of voluntary collective action in the fight against climate change was being undertaken with the launch of the campaign, which she noted, was an action the world must take today to preserve the climate for future generations (Maathai, 2003). She said that the target of 2007 would be achievable if one billion people out of the world's estimated population of six billion "dig a hole, put a tree in it and water it". The campaign is premised on the science of using trees as 'carbon sinks' whereby they absorb carbon dioxide and release it into the atmosphere as oxygen.

In Nigeria, government at various levels has in the past two decades embraced tree planting campaigns aimed at greening the environment, and checking desertification and erosion in most parts of the country. However, the campaign has degenerated into an annual fanfare without sustainable strategies to ensure that the campaign succeeds.

Lake Chad, a major water reservoir, has received attention regarding socio-economic and environmental implications in the event of its demise in the surrounding countries of Nigeria, Chad, Central Africa, and Cameroon. People of Sub-Saharan Africa now presume that the situations are amendable to the correction measures. There have been massive participatory rural appraisal and base line surveys, and environmental impact assessments, using the instrument of enlightenment, civic education and sensitization, as well as advocacy to relevant stakeholders for enabling law, workshops and capacity building. Through these ways, it had to a reasonable extent been brought to the fore all the dynamics of climate change as well as the whole challenges and ways to mitigate and adapt the action. Also, capacity building enables the rural communities and grass roots population to know the challenges of global warming and approaches to mitigation.

Liaisons with the ministries of planning are some of the approaches to work in concert with different states. Most of the states in Nigeria as well as other countries in Sub-Saharan Africa have started formulating environmental policies. An example is in Imo State of Nigeria, where policies have been formulated on the environment by agencies such as the Environmental Protection Agency

(ENTRACO), the Clean and Green Initiatives, Tree Planting and Transformation Agency in Rivers State of Nigeria, Agricultural Development Programs (ADP) in all the states of Nigeria, Land Reformation and Reforestation Agencies, National Orientation Agency, and the Nigerian Government 7-Point Agenda, etc. Traditional rulers, institutions of higher learning and research institutes, River Basin Development, Statistics and Planning, State and Federal legislators were also involved.

These agencies have been given a mandate to seriously tackle the challenges of climate change and embrace a value re-oriented approach. They have been put together and are collaborating with the programs and study of climate change adaptation and mitigation.

Target groups

The target groups had been rural farmers whose farmlands and fishing waters had been polluted and made poisonous; others were youths, children and other less privileged groups who had been held down by health problems caused by gas emissions and pollution. Subsistence rural farmers were also affected by highly reduced farm lands and numbers of seized economic trees, as well as storage problems brought about by extreme temperatures, which have affected agricultural production and habitation.

Based on the awareness campaigns carried out so far in this region, there have been provisions of alternative energy techniques, adopting of new agricultural processing and cottage industries, erosion control measures and improved agricultural land management practices, and establishment of standard statistical structures and data base for planning.

Global communities are in quest of identifying more farming practices, which reduce greenhouse gas emissions as well as act as resilient systems that are able to adapt to impacts of climate change. Organic agriculture is being considered as one of the appropriate farming systems that could serve the twin objectives of climate change mitigation and adaptation.

Adverse effects are likely to multiply if adaptation fails. This may then overstretch many societies, which may lead to destabilization and security risks, including loss of livelihoods, malnutrition and conflicts. In the light of these findings, the committee on World Food Security assessed the food situations in September, 2006, during their 32nd session and acknowledged that the World Food Summit´s target of halving the number of hungry people by 2015 will not be met; the number of undernourished has remained virtually unchanged since 1990–1992, although there has been a reduction in the percentage of under-nourished. Indeed, the current total global agricultural production is sufficient to feed the current world population, if the necessary technologies, multilateral environmental climate and conservative measures are put in place.

However, hunger, poverty and environmental degradation persist even as issues concerning global human security continue to increase. Moreover, the last decades provide uncompromising evidence of diminishing returns on grains despite the rapid increase of chemical pesticide and fertilizer applications resulting in less confidence that these high input technologies will provide for equitable household and national food security in the next decades.

The worldwide under-nourishment is not explained only by a lack of food as several causes of hardship lie outside the agricultural sector. However, there is need to seek new solutions to address the problems posed by a growing population and environmental degradation through new paradigms for agriculture and food supply chains (The organic agriculture paradigm). A number of researchers have proposed organic agriculture as a very important strategy for reducing the impact of climate change as it promotes improved soil quality and efficient water use, agro-ecosystems improvement and strong community knowledge processes, which in turn help to improve farm resilience against the adverse impacts of climate change as well as strengthen farmers' adaptive capacity. Soil fertility is maintained mainly through farm internal inputs (organic manures, legume production, crop rotation, tillage operations, etc.). Poor people who depend on agriculture are the most vulnerable to climate change. Increasing crop failures and livestock deaths are already imposing high economic losses and undermining food security. More droughts and increasing water scarcity may devastate large parts of the tropics and undermine irrigation and drinking water in the entire communities of already poor and vulnerable people.

In conclusion, integrated natural resources and environmental management strategies are moving towards the centre stage of climate change adaptation strategies in tropical landscapes. From the climate perspectives, this rationale is based on the premise that increased adaptive capacity is a first and most logical response to the complex threat of climate change, and that such capacity can only be attained through commitment to sustainable development. Organic agriculture is being practiced in Sub-Saharan Africa. International Organizations have indicated that it could ensure both greenhouse gas emissions reduction as well as development of resilient farming systems for adaptation and climate change related plans and mitigation strategies.

References

Abramovitz, J. N., 2001. Unnatural Disasters. *World watch*, Paper 158. Washington: World watch Institute.

Abutu, A., 2008. Mitigating Climate Change with TREES. The TIDE Newspaper, Wednesday, September 10, 2008.

Banuri, C., 2001. Setting the Stage: Climate Change and Sustainable Development. Chapter one in Climate Change 2001: *Mitigation. IPCC WG III Contribution to the Third Assessment Report.* IPCC, 8–15.

Chikaire, J., Nnadi, F. N., Nwakwasi, R. N. and F. C. Anaeto, 2009. Analysis of the Benefits of Organic Agriculture as Climate Change Adaptation Strategy in Ohaji Area of Imo State. *Proceedings of 5th National Conf. of Organic Agriculture Project in Tertiary Institutions in Nigeria.* Federal University of Technology, Owerri, Nigeria. 358–363.

FAO, 2007. Assessment of the World Food Security, CFS/2007/2. *Food and Agriculture Organization, Rome.* FGN, 2009. Official Gazette. Vol.1, No. 2. *Federal Government of Nigeria,* Abuja.

Lobell, D. B., Burke, M. B., Tebald, C., Mastrandrea, M. D., Falcon, W. P. and R. L. Naylon, 2008. Prioritize Climate Change Adaptation Needs for FOOD Security in 2030. *Science.* 319; 607–610.

Maathai, W. 2003. Green Belt Movement. Unbowed: A Memoir. The Challenge for Africa, and Replenishing the Earth. Sharing the Approach and the Experience. *Lantern Books.* ISBN 1-59056-040.

McClean, C. J., 2005. African Plant Diversity and Climate Change. *Annals of the Missouri Botanical Garden*, 92(2):139- 152.

Morgan, R. P. C., 1986. Soil erosion and conservation. *Applied Geogr. Series.* 1990: 1–13.

Munasinghe, M., 2000. Development, Equity and Sustainability (DES) in the context of climate change and its linkages with development, equity, and sustainability: *Proceedings of the IPCC.*

Noss, R. F. 2001. Beyond Kyoto Forest Management in a time of Rapid Climate Change. *Conservation Biology.* 15:3. 578–590.

Okoye, I. G., 2008. Climate Change and Attendant Implications to Agricultural and Rural Development. *Workshop on Environmental Issues* In: *Agricultural/ Agro-Industrial and Rural Development Projects.* Agricultural and Rural Management Training Institute (ARMTI). Ibadan, Nigeria. P. 15.

RIOA, 2008. Organic Agriculture and Climate Change: A Monograph by UNCTAD and FIBL. *Research Institute of Organic Agriculture, Bonn, Germany.*

Smil, V., 1993. Global Ecology: Environmental Change and Social Flexibility, London. 240 pp.

UNDP, 2002. Population, Environment and Development. In: *The Punch* (28/8/02). Experts scaling back estimates of World Population growth.

World Bank, 1988. Poverty and Hunger. Issues and options for food security in Developing Countries. Washington, DC.

World Conservation Monitoring Centre, 1992. Global Biodiversity. Status of the Earth's living Resources. Cambridge, UK, WCMC.

Amidu Owolabi Ayeni
University of Lagos, Nigeria

Rainfall variabilities in Sub-Sahara Africa: Oberservations and implications in Ilorin, Southwest Nigeria

Abstract

This paper examines the effects of global climate change on annual rainfall intensity in Ilorin, Southwest Nigeria. Rainfall data of 60 years (1946–2006) were collected and analyzed. Arithmetic mean and trend analysis were employed to generate monthly mean from the daily readings to determine the long-term dispersion scenario observed from the rainfall data over the years. The result shows that the start, annual intensity and end of rainfall over Ilorin did not differ while the length and amount of the rain shows variation over the area. The rain starts as between 7th March and 4th April. It was also observed that the end of rain had a late retreat in the 1940s, 1950s and 1960s, while 1970s, 1980s, 1990s and 2000s experienced early retreat of rainfall. It can, therefore, be concluded that the start, end and amount of rain experienced little variation, but the length of rain exhibited variation and this could influence the use of the rain for agriculture and for other purposes.

Introduction

A thorough understanding of climate change requires a working knowledge of climate systems. Climate involves principally the atmosphere, hydrosphere, lithosphere, biosphere, and the cryosphere. Humans have always struggled to cope with climate variability. Climate variability has been observed to have increased the number of floods and droughts in recent decades, and equally, climate change has also emerged as one of the key environmental issues of concern over 30–100 years (Enete and Amusa, 2010). However, handling the impacts of climate variability and change as well as uncertainty of climate sensitivity, which have hardly changed in over 20 years, is still a challenge to human society (Awake, 2008). There are various elements working together to form the totality of weather and climate in which precipitation and temperature wind are the most salient elements that change rapidly with time and space.

Precipitation, however, has been known to be one of the various forms in which water falls from the sky. The most frequent forms of precipitation are rain, snow and hail. The distribution varies from one geographical region to the other. Nonetheless, if each region in the world would receive 860mm of precipitation, the total annual precipitation would be evenly distributed (Olaniran, 1990). But as it is, annual distributions of precipitation range spatially from less than 10mm per month to more than 300mm per month depending on the location (Pidwirny, 2006). This is due to the variability and fluctuation in the annual precipitation received over various regions of the world.

In recent times, examination of various precipitation data indicates a run of dry years for the Sub-Sahara region dating back to the 1940s (Bryson, 1973). Lamb (1985), in particular noted that the years 1942, 1949 and 1968–1983 were drier than normal. Adeyemi (2000a) observed dry periods of varying intensity and magnitudes in the south of Sub-Sahara Africa. Such trends, variability and fluctuation of climate seem to be of interest to global climate change studies because of their diverse effects, particularly drought and flood (Ayeni *et al.*, 2011), yet, it is the most uncertain aspect of meteorological research. The realization of extreme changes in climate with time only came to the limelight when indisputable evidence was obtained for ice ages in the 1840's, although the climate has altered sufficiently yet in many parts of the world within the last few thousand years (Awake 2008). Unfortunately, reliable weather and climate data have only been kept during the last hundred years (Haile, 2005) and therefore, it is only the recent climatic fluctuations that can be investigated adequately (Rahmstorf *et al.*, 2007; Rahmstorf *et al.*, 2012). In spite of man's remarkable advances in science and technology, his economic and social well-being still depend on weather and climate and any anomaly in these two variables will not only affect man but the totality of man's environment.

It is on these facts that the daily rainfall measurement that allows the assessment of rainfall variability between 1946 and 2007 forms the basis of this study in which the temporal distribution of rainfall over Ilorin, Southwestern Nigeria, was examined for variation in rainfall amounts and its attributes such as trends of the start, end and length of rainfall, as well as amount and intensity of rainfall received over the area within the stipulated period. The objectives of this research are (i) to determine the trend of the start, end and length, and mean intensity and of rainfall received over the area within a 60-year (1946–2007) period and (ii) to determine the magnitude of the amount of rainfall received over the area.

The study area

The study area is located in the northern part of Kwara State and ranges between latitude 8°24' to 8°32'N and longitude 4°29' to 4°35'E (Fig. 1). According to the Nigeria's National Bureau of Statistics (NBS, 2007), the population of the area in 2006 was 182,791. The topography is generally undulating with some noticeable hills such as Sobi and others. Average elevation is about 300m above mean sea level. The main river draining the area is Asa River, which occupies a fairly wide valley and flows in the south-north direction. The geological formation is basically crystalline rocks of the Precambrian age.

Fig.1: Kwara State Showing the Study Area (Ilorin West).

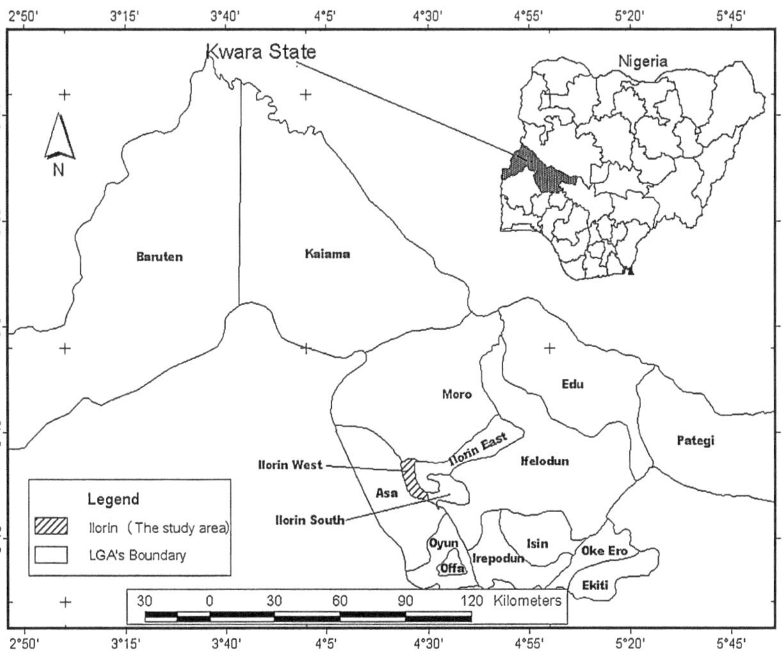

There are two air masses blowing over the area: the dry north easterlies and the moist south westerly. The result is the emergence of two well-marked seasons: the dry and wet seasons. The dry season is experienced from November through

March, while the wet season is experienced from April through October. The region experiences a rainfall with an annual range of 1000–1250mm (Barbour *et al.*, 1982; Uluocha and Ekop, 2002). Temperatures are constantly medium to high between 26°C to 28°C. The vegetation is predominantly Guinea savanna, which degenerates with increasing aridity (Barbour *et al*, 1982; Uluocha and Ekop, 2002). The economic activities within the area are mostly agriculture-based and trading.

Methodology

The daily rainfall data (1961–2007) used in this study was sourced from Nigerian Meteorological Agency (NIMET) – Ilorin Airport Meteorological Unit and Lower Niger River Basin Development Authority (LNRBDA). The data was analyzed using Microsoft Excel 2007, while simple arithmetic means were calculated from daily readings to generate monthly and annual data. Trend analysis was performed (using Microsoft Excel 2007) to determine the long-term variations in observed rainfall data for start, end and length of rain over the studied number of years. The arithmetic mean formula is given thus:

$$X = \frac{\Sigma x}{N}$$

Where;

X = Arithmetic mean
Σx = Sum total of the x^{i-n}
N = number of items $^{i-n}$ in the distribution

Z score also employed to facilitate comparisons of mean deviations (index) of amount and intensity of rain between 1946 and 2006. The Z score formula is given thus:

$$Z_x = \frac{x - \mu_x}{\delta_x}$$

Where;

Zx = Standard or normal distribution
x = individual mean
$\mu x = \Sigma(x)$ = the mean of the population
δx = the standard deviation of the probability distribution of x

Findings

The findings reveal that the periods of years between 1946–1949 and 1971–1975 had each predominantly consistent starts of rainy season and subsequently experienced more wetness (Fig. 2). The year 1993 and 2005 experienced an early start of rain. On the other hand, the years between 1950 and 1956 experienced a consistent lateness in the start of rainy season with 1955, 1965 and 1988 experiencing the highest delay in the start of rain. Generally, the early and late start of rainy season between the periods of 1946–2004 in Ilorin seemed to be almost equal, depicting that the area was not a predominantly wet region.

Fig. 2: Day of start of rain, end and length form January 1st.

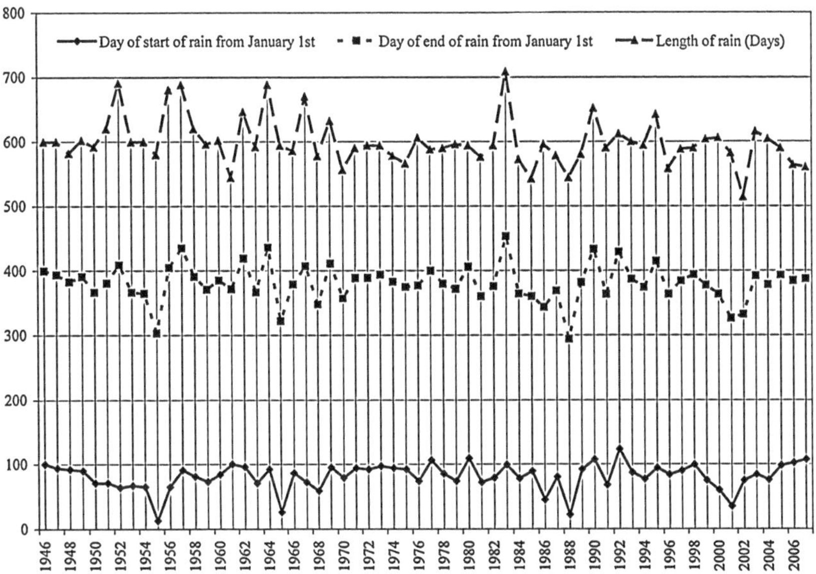

The years 1961, 1985, 1988 had similar early end of rainy season while 1983 on the other hand had a late end of rainy season (Fig. 2). Ilorin region had early ends of rainy season between 1946–2006 with few exceptions during 1952, 1956, 1957, 1962, 1964, 1967, 1983, 1990, 1995 and 2007. It could be concluded that more years between the periods of 1946–2006 experienced dryness due to the early ends or retreats of rainy seasons. There were little fluctuations in the length of rainy season, with years 1952 and 1961 that experienced the longest and shortest length respectively. The period between 1950 and1959 experienced a consistent long length of rain while the period between 1970 and 1975 experienced short length of rain (Fig. 2). The highest amount of rainfall was received in the year 1968, while years 1961 and 1989 had the lowest amounts. Between 1949 and 1957, a consistently high amount of rainfall was experienced depicting wetness years; while on the other hand, the years between 1974 and 1979 experienced a consistent low amount of rainfall depicting dryness (Fig. 3).

Fig. 3: Index of amount and intensity of rain between 1946 and 2006.

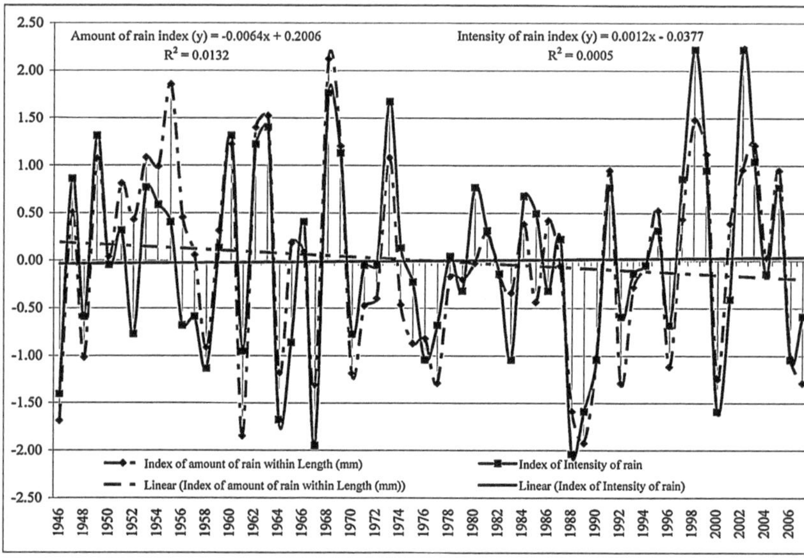

The intensity of rainfall shows a high variability, the years 1968, 1973 and 1998 experiencing high intensity of rainfall with 1998 being the highest. The years 1964, 1988, 2000 and 2006 experienced quite low intensity of rainfall with 1988 being the lowest intensity of rainfall (Fig. 3). The analysis revealed that the annual intensity of rainfall over Ilorin did not differ over the period studied. The start of rain did not differ with the rains starting as early as 7th March and as late as 4th April. The end of rain did not show much difference either with its decadal retreat falling between late October and early November and its mean annual retreat in 28th of October. The length of rain received on the other hand shows variation in the mean of the **decades**, while the amount of rain received followed the same pattern showing the little or no variation.

It was also observed that the end of rains had a late retreat in the 1940s, 1950s and 1960s, while the 1970s, 1980s, 1990s and 2000s experienced early retreat of rainfall. In other words, it can be deduced that as years went by, rainfalls retreated earlier. This also means that the rainfall pattern in Ilorin seems to be fluctuating between wetness and dryness based on this study. It can therefore be concluded, that the start, end and amount of rain experienced little variation, but the length of rain exhibited variation and this could influence the efficiency of rain.

Environmental threats in a climate change scenario

Expected environmental threats in a climate change scenario can be summarized under the following:

- Threat to weather: Increased earth temperatures lead to fast cycling of water, which leads to severe storms and flooding in some areas and drought in others. Extremes of wind are also expected.
- Threat to the ecosystems: Global change may put the forests under stress by attracting acid rain and increasing the frequency of forest fires and species diversity reduction.
- Threat to human health: Increase incidences of ultra-violet rays resulting from the depletion of the ozone layer may lead to skin cancer and sun burns. Increased temperatures due to increasing greenhouse gases may lead to heat related illnesses. There will also be a change in the spatial pattern of diseases, for instance, malaria occurs more in the tropics, because *Anopheles* mosquito cannot survive cold weather.
- Threat to food and water: There may be scarcity of food resulting from unpredictability of global natural resources, notably rainfall. Farming may lead to environmental refugees. Marine productivity is also at a risk. However, nutrient-fixing may be hampered due to reduced vertical soil moisture distribution. With sea level rising, coastal farmlands may be inundated and destroyed. Also, the rate of the metabolism is expected to increase in warmer water, leading to death of marine animals and contamination of seafood.

Possible solutions

Although, past emissions of greenhouse gases will remain in the atmosphere for decades or more to come, urgent reduction of today´s emissions from cars, power plants, land use, and other sources will determine the kind of world the next and/or future generation will live in. Therefore, the solutions to the threatening climate change lies in hand of governments and individuals, and businesses which are expected to assist by adopting various options for reducing greenhouse gas emissions. This can be achieved by:

- Increasing fuel-efficient vehicles and higher efficiency appliances.
- Establishing strategies/measures to protect existing forests and restoring depleted ones, and
- Eradicating provisions that promote the use of coal and oil resources that are harmful to the environment.

Individuals can reduce the need for fossil fuels and often save money by using energy-efficient appliances and less electricity in general, driving less and driving more fuel-efficient and less-polluting cars, and insulating their homes. Other measures include increasing the efficiency of energy saving appliances and encouraging consumers to adopt other efficiency measures (www.wecansolveit.org).

Conclusions

Given the fact that the nature of the system is so much a complex one, the future effects are difficult to predict. There are many speculations as to result in cynicism. However, we should expect a lot of surprises. Drier areas may become either wetter or even drier depending on their locations and vice versa. In some areas, the changes may be self re-enforcing, with human beings having to cope within, while in some, it may be self-canceling. With the nature and complexity of global climate change coupled with rapidly increasing population, future impacts of man on climate may become more difficult to predict. The consequences are many and in most cases injurious to man and his environment.

Acknowledgments

The author would like to acknowledge the Nigerian Meteorological Agency (NIMET) and Lower Niger River Basin Development Authority (LNRBDA), Ilorin, for the rainfall data.

References

Adeyemi, A. S., 2000a. "Climate Change and Environmental Threats", in: Jimoh, H. I. and I. P. Ifabiyi (eds.), Contemporary Issues in Environmental Studies, Haytee Publishing Ltd., 309 pp.

Ayeni, A. O., Soneye, A. S. O. Fasunwon, O. O., Miteku, R. T. and L. A. Djiotang-Tchotchou, 2011. Water Resources Development Optimization in a Climate Change Scenario: Case Study of Benin-Owena Basin, Nigeria. Research Journal of Environmental Sciences, 5: 59–67.

Awake, 2001. "Cherrapunji – One of the Wettest Places on Earth," in Awake! Pennsylvania, Watch Tower Bible and Tract Society. May 8, 2001, 22–23.

Awake, 2008. Global Warming, Is Planet Earth in Peril? In: Awake, May 2008, Pennsylvania, Watch Tower Bible and Tract Society, pp. 3–9.

Barbour, K. M., Oguntoyinbo, J. S., Onyemelukwe, J. O. C. and J. C. Nwafor, 1982. Nigeria in Maps, London, Hodder and Stoughton. 209 pp.

Bryson, R. A., 1973. Drought in Sahelia: Who or what is to blame, Ecologist 3(10): 360–371.

Enete, A. A. and T. A. Amusa, 2010. Challenges of Agricultural Adaptation to Climate Change in Nigeria: a Synthesis from the Literature, Field Actions Science Reports [Online], Vol. 4, 2010, http://factsreports.revues.org/678.

Haile, M., 2005. Weather patterns, food security and humanitarian response in Sub-Saharan Africa. Phil. Trans. R. Soc. B, 360: 2169–2182. doi:10.1098/rstb.2005.1746.

Lamb, P. J., 1985. Rainfall in Sub-Saharan West Africa during 1941–83. Z. Gletsch. Glaziol. Geol., 21, 131–139.

National Bureau of Statistics: NBS 2007. Official Gazette (FGP 71/52007/2,500(OL24): Legal Notice on Publication of the Details of the Breakdown of the National and State Provisional Totals 2006 Census, National Bureau of Statistics (NBS) Publication. Federal Republic of Nigeria, www.nigerianstat.gov.ng2007.

Olaniran, O. J., 1990. Changing Pattern of Rain days in Nigeria, Geo-journal, 22 (1): 99–107.

Pidwirny, M., 2006. "Fundamentals of Physical Geography", 2nd Edition. http://www.physicalgeography.net/physgeoglos/c.html.

Rahmstorf, S., Cazenave, J. A., Church, J. E., Hansen, R. F., Parker, Keeling D. E. and R. C. J. Somerville, 2007. Recent Climate Observations Compared to Projections. Science 316: 709, doi:10.1126/science.1136843.

Rahmstorf, S., Foster, G. and A. Cazenave, 2012. Comparing climate projections to observations up to 2011. Environ. Res. Lett. 7: p. 5, doi:10.1088/1748-9326/7/4/044035.

Uluocha, N. O. and G. Ekop, 2002. Nigeria: Geography. In: Osuntokun, A. Aworawo and F. Masajuwa (eds.), History and Cultures of Nigeria up to AD 2000, Lagos, Frankad Publishers, pp. 3–19. http://www.wecansolveit.org/content/pages/60? (26/08/2008). I want to be part of the solution.

C. U. Oyegun
University of Lagos, Nigeria

Global climate change impacts on Sub-Sahara Africa: The case of Nigeria's shorelines

Introduction

Climate refers to the average weather condition and its variability over time and space. Climate varies from place to place depending on latitude, distance from the sea, vegetation and terrain characteristics. Climate also varies in time from season to season, year to year, decade to decade and on longer time scales which extend to millions of years. The Inter-governmental Panel on Climate Change (IPCC, 1996) defined climate change as statistically significant variation in the mean state of the climate or of its variability which typically persists for decades or longer periods of time. Climate is predictable and fairly stable over large continental and global scales even though extreme atmospheric conditions also influence the weather of a place. The elements of weather and climate which influence our everyday life include average temperature, minimum and maximum temperature, wind near the earth's surface, various types of precipitation, humidity, cloud cover and solar radiation. Others include the vertical structure of the atmosphere, the influence of the underlying land and sea, atmospheric circulation and its interaction with ocean currents, the albedo of the earth's surface, vegetation and soil moisture. In addition to these, the climate of the world is influenced by other factors that control the radiative balance, like atmospheric composition, solar radiation and volcanic eruptions. There is the need now more than ever before to understand the climate system which includes the dynamics and composition of the atmosphere, the oceans, the ice and snow cover, the land surface and its features, the world of living organisms and the mutual interactions within and between these spheres (Oyegun and Ologunorisa, 2002). In the light of the foregoing, some questions are pertinent: Is climate change a fact or fiction? Has global climate remained constant since the beginning of the earth? What are greenhouse gases? In what ways have the activities of man impacted the climate of the globe? What are the observations, problems and solutions of climate change in Sub-Saharan Africa? This paper attempts to address the above questions particularly as they affect the 40.000km^2 shore-zone of Nigeria.

First, variations within the earth's atmosphere and processes in other parts of the earth like in the oceans and ice-caps including the effects of human activities can bring about climate change. These external factors which shape the climate of a place are referred to as climate forcing and include processes like variations in solar radiation, the earth's orbit and greenhouse gas concentrations. Glaciers have been recognized as one of the most significant indicators of climate change, because they advance substantially during the cooling phase of the climate, for example during the little ice age and retreat during phases of climate warming on moderate time scales. Over the last 100 years, the reality all over the globe is that glaciers have been unable to regenerate enough ice during winter to make up for losses in summer. According to Miller (2004), the past 900,000 years of the 4.7 billion years of earth history have witnessed prolonged periods of global cooling and warming. Thick layers of glacial ice covered vast areas of the earth's surface for about 100,000 years during the cold or glacial periods in which there was a significant decrease in atmospheric temperature. These glacial periods alternate with warmer interglacial periods, which last for between 10,000 and 12,500 years with significant increases in the temperature of the atmosphere. The world is presently passing through an interglacial phase and the past 10,000 years before present have witnessed a stable atmospheric temperature of 15°C even though the temperature of the earth's atmosphere has been on the increase since the 1980s. Houghton and Woodwell (1989) describe the present condition of the earth in the following statement:

> *The world is warming. Climatic zones are shifting. Glaciers are melting. Sea level is rising. These are not hypothetical events from a science fiction movie; these changes and others are already taking place and we expect them to accelerate over the next years as the amounts of carbon dioxide, methane and other trace gases accumulating in the atmosphere through human activities increase.*

Climate change also results from the interaction between the atmosphere and the oceans of the world over a longer time scale. Many climatic fluctuations of the recent past like the El Nino Southern Oscillations, the Pacific Decadal Oscillation, the North Atlantic Oscillation and the Artie Oscillation owe their occurrence mainly to the different ways heat is stored in the oceans and its motion between the different reservoirs. Over longer time scales, thermohaline circulation which is the movement of ocean waters over temperature and salinity gradients play a key role in redistributing heat, thus bringing about dramatic changes in climate. There are also non-climate factors driving climate change. Recent studies have shown that the radiative forcing of greenhouse gases is the primary cause of modern changes in global climate.

Earth's paleo-climate, human activities and green house gas emission

It is now common knowledge that climate change has been an integral part of the history of the earth and that long-term global climate change is inevitable. There is nothing that the global society of man can do about it, because it spreads over hundreds and thousands of years. On the other hand, man must seek to understand and manage short-term changes in climate, which influence global temperatures within the life span of present generations. Christopherson (2002) affirms that the last two decades were dominated by record high global temperatures for both land and ocean and for both day and night with 1998 recording the highest temperatures since the past 420,000 years. The emission of greenhouse gases particularly carbon dioxide into the atmosphere through the agency of human activities is mainly to blame for the current warming trend. According to IPCC (2007), the atmospheric concentration of carbon dioxide ($C0_2$) in 2005 was 379ppm compared to the pre-industrial levels of 280ppm. Table 1 below shows the carbon dioxide concentration of the lower atmosphere between 1774 and 2050.

Table 1. Carbon dioxide (CO_2) concentration in the lower atmosphere

Year	CO_2 concentration (%)	Parts per million (ppm)
1774	0.028	280
1888	0.029	290
1970	0.032	320
1985	0.035	350
2000	0.037	370
2020 (estimate)	0.055	550
2050 (estimate)	0.060	600

Source: Christopherson (2002).

The picture which emerges here is that the percentage concentration of carbonite in the lower atmosphere has been increasing progressively since the dawn of the industrial revolution, which began in the 1750s. The more carbon dioxide and the other radiative gases are in the atmosphere, the more the air temperature increases. These increases are estimated to reach 560ppm before the end of the 21st century. Evidence in the literature shows that CO_2 is substantially higher now than it has been at any other time in the last 800,000 years. The human activities of the modern period that have contributed to the significant amounts of greenhouse gases in the troposphere include:

i) The sharp rise in the use of fossil fuels that release large amounts of greenhouse gases viz. carbon dioxide (CO_2) and methane (CH_4) into the atmosphere.
ii) Aerosols – particulate matter of sulphur from fossil fuel combustion in the atmosphere which exert a cooling effect on weather and climate.
iii) Cement manufacture – 3rd largest cause of man-made CO_2 emission.
iv) Land use arising from urbanization and industrialization.
v) Ozone depletion of the stratosphere.
vi) Animal husbandry – which accounts for 18% of world's greenhouse emission.
vii) Deforestation, clearing and burning of grasslands and forests to grow crops which release carbon dioxide (CO_2) and noxious gases into the atmosphere.
viii) Cultivation of rice in paddies and the use of inorganic fertilizers that release nitrous oxide (N_2O) into the atmosphere.

Miller (2004) reported that the two largest contributors to CO_2 emission are the world's thousands of coal-burning power and industrial plants and more than 700 million gasoline-burning motor vehicles and that the emission of carbon dioxide from coal-burning power and industrial plants in the US alone exceeds the combined carbon dioxide emission of 146 nations where 75% of the world's population lives. Again, Christopherson (2002) posits that carbon dioxide alone is responsible for 64% of the global warming trend and that through 1999 carbon dioxide levels were more than 30% higher than they were in the pre-industrial period. He further states that the global emission of carbon from fossil fuels reached 6.3 billion metric tons in 1999 and when combined with oxygen in carbon dioxide gas, it increased 18.9 billion metric tons. This is 400% more than the 1.6 billion metric tons of carbon in 1950. Hence, carbon dioxide is increasing in concentration at the rate of 0.5% per year. Table 2 below shows that whereas the carbon dioxide emission of developed countries attained 73% of total world emission in 1995, the United States of America alone contributed 23% of this greenhouse gas emission. Comparatively, the developing nations together contributed 27% of the world's total greenhouse gas emission with China singularly contributing 14% to this value. In the 21st century, the developing countries are identified as the sector of the world with the greatest probable growth in fossil fuel consumption and new carbon dioxide production.

Table 2. Origin of excessive carbon dioxide (CO_2)

Status of Countries	**1995 emission of CO_2**	**2025 emission of CO_2 (Projected)**
Developed	73%	50%
	(USA 23%)	(USA 17%)
Developing	27%	50%
	(China 14%)	(China 15%)

Source: United Nations Environmental Programme (UNEP, 1995).

Of the developed countries of the world, Australia and the European Union countries record the highest greenhouse emission levels. The 2002 report of the Australia Institute shows that the country had the highest greenhouse gas emission per person in the year 2001 mostly as a result of land clearing. The country's emission per person was double the average for all industrialized countries and 35% higher than that of the United States, which had the second highest per capita emission of greenhouse gases in the world for the year under reference. China, Brazil and India are the fastest growing economies in the third world. The projected emission level of 50% for developing countries for the year 2025 can only be reduced if the governments of third world countries make concerted efforts to reduce emission levels through the use of renewable, low temperature and labor intensive energy sources. The industrialized countries also must drastically cut emission levels concurrently for the projected levels to be attained. Ruddiman (2003, 2005) and Ruddiman *et al.*, (2005) have formulated the early anthropocene hypothesis which suggests that the rise of agriculture and the accompanying deforestation during the period 5,000–8,000 years before present led to the increases in carbon dioxide (CO_2) and methane (CH_4) concentrations in the atmosphere. They further argue that these increases which reversed previous declines may have been responsible for delaying the onset of the next glacial period. This speculation is yet to be substantiated. Methane (CH_4) as a radiative gas contributes to the overall greenhouse effect of the atmosphere with an annual increase of 1%. The analysis of trapped air bubbles in ice with ages of over 500 to 27,000 years before present show that the concentration of methane in the past was about 0.7ppm. Current atmospheric concentration of the gas is about 1.73ppm, which is more than double its pre-industrial levels.

Immediate Impacts on Nigeria's Shorelines

The immediate impact of climate change in Nigeria's shorelines, as elsewhere in Sub-Saharan Africa region, is fluctuation in rainfall patterns with dry areas becoming drier and wet areas becoming wetter. Sea level is rising, coastal erosion is on the increase and low-lying coastal communities are being submerged under water as highlighted below. The problems of Nigeria's coastal zone are numerous. They include coastal/riverbank erosion accentuated by local subsidence, flooding, sedimentation and siltation of the water ways.

Table 3. Erosion rates along Nigeria's coastline

Location	Rate of Erosion per annum
Badagry Beach (Lagos State)	2–6m
Victoria Beach (Lagos State)	25–30m
Awoye/Molume (Ondo State)	20–39m
Ogborodo/Escravos (Delta State)	18–24m
Forcados Beach (Delta State)	16.8–22m
Brass (Bayelsa State)	16–19m
Ibeno – Eket (Akwa – Ibom State)	10–13m

Source: Oyegun (2007).

Table 3 shows the rates of coastal erosion along Nigeria's coastline. The reasons advanced for coastal erosion globally include: the worldwide rise in sea-level arising from global warming, the fact that since about 5,000–7,000 years before present, coastal dynamics have attained a state of equilibrium resulting in the drastic reduction of the materials washed onshore from the sea bed and from sea cliff erosion; climatic oscillations and the influence of man.

Coastal and river bank erosion

The 800 km coastline of Nigeria is witnessing accelerated erosion at some sections while active accretion is dominant at other stretches of the coastline (Oyegun, 1989). The Victoria Beach in Lagos, the 50–75 km long transgressive mud coast between the Benin River mouth and the Lekki Lagoon complex and the mouths of the major drainage arteries of the Nigerian landscape are witnessing active coastal erosion. At the other end of the continuum, sections of the arcuate Niger Delta and protected reaches of shoreline are actively accreting. The problem of riverbank erosion at the mouths of the main drainage arteries of Nigeria's land space arises from the impact of breaking swell waves on the unconsolidated materials of the riverbanks. The major river systems in Nigeria include the Forcados River, River Nun, Brass River, the New Calabar River, Escravos River, and the Benin River. The impact of waves generated by the propellers of river crafts on the banks of these rivers alongside the effect of river currents and the reversible tidal currents of the semi-diurnal tidal regime also need to be mentioned. Abam (1988) affirms that this problem of river bank erosion is predominant along the outer banks of rivers with gully heads that retreat at 0.4 m per year in the creeks of the Niger Delta wetlands. A report of the Research Planning Institute (RPI,1985) concluded that acute coastal erosion is taking place along the transgressive mud coast northwest of the Benin River mouth. The acute erosion at this coastline has resulted in the formation of a Berm cliff that is over 2 m in height, cut in mud, and fronted by an off-shore terrace of mud which extends for 1–2 km offshore. Ebisemiju (1988) reported that based on the analysis of aerial photographs taken in this area in 1963, 1973 and 1981, and from field measurements, shoreline retreat ranges from 20–39 m per annum. He further stated that the recession of the shoreline at this coastal reach has resulted in the loss of 487 ha of coastal land, a gross top soil loss of over 2.5 million m^3, and a reduction of the coastal plain by as much as half of its former width in only eight years. The encroaching saline water has led to the contamination of fresh water sources, resulting in long distance treks inland in search of potable water. There is also the death of freshwater water vegetation as well as the demise of vast resources of the mangrove swamp, which constitutes the hatchery for several varieties of fish and other forms of marine life. These have combined to deplete the rich fish resources of the region, because the habitat is no longer conducive to previous colonization rates of fish for spawning and breeding. This has forced many of the younger people to leave the area and move to other coastal locations to make a living from fishing. It is suggested that the withdrawal of oil and gas from the Agbada formation is accelerating local subsidence and hence the landward translation of sea level. The net effect of this is accelerated coastal erosion. The essentially low-lying nature of the terrain of the area is a contributory factor to the erosion problem with the most affected settlements being Awoye, Idogo, Ayetoro and Erunama, which are mainly fishing villages. The increasing level of the sea due to global warming has further exacerbated the hardship of the people of the region, because shoreline erosion has intensified in the recent past. Lawson (1988) reported that the Federal Ministry of Works and Housing in appreciation of the enormity of the problem has undertaken sand replenishment projects at the following locations:

Victoria Beach, Lagos, involving 2.5 million m^3 of sand; Bomadi, Patani, Burutu and Ogulagha, 4 million m^3 of sand each; and Opobo, Brass, Bonny, Sangana, Ndoni, Akassa, Omoku, 3.8 million m^3 of sand.

King (1987) warned that sand for beach nourishment should be dredged off-shore beyond the limits where wave action entrains bottom sediments, so that the delicate equilibrium of the immediate off-shore area is not disturbed. The Niger Delta Basin Development Authority, a federal government agency based in Port Harcourt, undertook erosion protection works at Opobo in 1981 at a cost of N11.3 million. The Federal Ministry of Works has completed similar projects for Brass at a cost of N5.2 million, Oron and Queenstown at a cost of over N5 million, while Ibe and Antia (1983) have put the cost of coastal stabilization works in Nigeria at over N30 million per annum. These statistics show government's obvious concern with the erosion problem. However, the approach has been at best "ad hoc", because these attempts at solving the problem are neither coordinated nor based on an understanding of the dynamics involved. The maintenance of these structures has also not been consistent resulting in the current disrepair of many of the engineering works. The eustatic rise in sea level consequent on climate change is accelerating the rates of coastal erosion unless urgent steps are taken to salvage settlements in the coastal belt of Nigeria.

Depletion of Victoria Beach and Forcados South-Point

The barrier-lagoon coastline of the Lagos – Lekki area evolved from the coalescence of fossil barrier sand ridges, which sandwich the lagoons that run parallel to the coastline. Littoral drift direction at this section of the coastline is from west to east with sediment transport estimated at between 500,000 m^3 and 2 million m^3 of sand per annum (Nedeco, 1961). The need for a harbor in Lagos resulted in the cutting of an access route across the barrier ridges to reach the more sheltered water of the Lagos lagoon. The beginning of construction work on the protective moles of the Lagos Harbor about the year 1911 was also the beginning of acute beach erosion at Victoria Beach, Lagos. Usoro (1971) reported about mean beach retreat rates of 23 m per year at Victoria Beach, Lagos. Ibe and Antia (1983) reported retreat rates of 69 m per year for the same beach. Presently, the swell waves of the Atlantic Ocean are breaking less than 10 m from the road nearest to the ocean in Victoria Island. If this trend is left unchecked, the whole of Victoria Island comes under the threat of the ocean. The reason for the accelerated erosion at Victoria Beach is not unconnected with the obstruction of sand movement across the entrance of the Lagos Harbor. This deprivation of the free flow of the vast sediment drift identified above has resulted in the accretion of Light House Beach on the up drift area and acute erosion of Victoria Beach on the down drift section which has lost about 12 km^2 of beach space to the Atlantic Ocean since 1911.

Oyegun (1991b) reported that between 1960 and 1988 there was a cumulative land loss of 13.46 km^2 at Forcados South Point, Youbebe to the Atlantic Ocean. This is in spite of the series of conservation strategies embarked upon by Shell Petroleum Development Company (SPDC), Western Division, to save this small but precious piece of land on which were located eight oil wells of the company. The most serious erosion zone at Forcados South Point and Youbebe has been protected with concrete structures and synthetic longard tubes emplaced perpendicularly to the shoreline to trap sand movement alongshore. By February 1987, at least three of these longard tubes had been washed away due to the impact of breaking waves. These tubes measure between 30 and 50 m in length and the reality on ground has shown that they are not long and sturdy enough to withstand the energy of breaking waves at the southern estuary of the Forcados River. Possibly, the use of rock boulders similar to the ones used for the protection of the entrance into Lagos Harbor would appreciably ameliorate the problem. There is, however, the need to extend the defense system to Ogulagha, because the materials eroded from Youbebe area are carried by the littoral drift and deposited across the entrance to Ogulagha by sea. All the coastal settlements which situate along the beach lie directly in the path of breaking swell waves, which possess as much as 1.24×10^5 kJ of wave energy at the peak of the wet season at Forcados South Point (Oyegun, 1991). The wave energy of the swell waves which wash the coastline of Nigeria will progressively increase with the rising sea level and thus enable bigger waves to break closer to the shore and result in more acute erosion.

Floods

The problem of flooding in the coastal ecological belt of Nigeria results from manmade and natural causes. The high population density of the highly built-up Lagos area has induced local subsidence of the largely reclaimed Lagos Metropolis. This combines with the inadequate drainage facilities for storm runoff to flood vast sections of the metropolis with almost every rainfall. The need to pump drain sections of Victoria Island, Ikoyi, Apapa and sections of the Lagos mainland has become very glaring with the recent spate of urban floods in the area. Land degradation in the arcuate Niger Delta is mainly in the form of floods given the low-lying terrain of its backshore limits. The floods are in part caused by the semi diurnal tidal regime of the area and more fundamentally by the climate factor. Fubara (2007) reported that mean annual rainfall within the Niger Delta ranges from 2.500 mm in its northern sections through 3.000 mm in the middle of the Delta to over 4.500 mm in the coastal towns of Opobo, Bonny and Brass. The highest flood levels occur in the months of September/October of each year's wet season. The tidal flats of the Niger Delta generally lie below 2.5 m above sea level hence over 700.000 ha of land area is under water at high tide in the region. This implies that as much as 45% of the Niger Delta region is inundated daily at every flood tide, which fluctuates with the ebb tide every 6 hours in the semi-diurnal tidal regime. The only lasting solution to the problem would derive from a massive and comprehensive land reclamation scheme in which polders are used to restrict water to creeks and canals. This would result in massive resettlement schemes to make way for the construction of massive water reservoirs in the central portions of the Niger Delta to control water levels and check the massive seasonal floods in the region.

Sedimentation and siltation of waterways

Sedimentation and siltation of the distributaries of the Niger Delta have been the major reason for the frequent dredging of the waterways of the region because the zone is a sedimentary environment. This factor also explains the decline in the use of the early ports of Forcados and Burutu as viable ports in the Niger Delta. Ibe and Antia (1983) have pointed to the existence of river mouth bars at the points where the Forcados River, the Brass River, the New Calabar River and to a lesser extent the Benin and Escravos Rivers enter into the ocean. These river mouth bars are shallow sub-marine features, which precipitate the breaking of waves thus translating the immense wave energy to impact the unconsolidated clastic particles of the estuary banks. This phenomenon brings about serious erosion problems in the immediate environs of the riverbanks in the direction of the breaking swell waves. The Forcados estuary is becoming shallower with time and rapidly. Pugh (1954) reported that the sand bar at the mouth of the river has silted up from 6.4m to 4m in 50 years. This channel alluviation is in part due to the accumulation of sediment discharged into the Forcados drainage system. At Ogulagha, situated along the southern estuary of the Forcados River, a period of 3 months without dredging was sufficient to bring about the growth of sand spit measuring 450m long, 65m wide and 1.3 m in height above the estuary surface at low tide, thereby blocking the access to the settlement's jetty. This accumulation of sediments at the mouth of the canal created navigation problems for river vessels that belong to the Shell Petroleum Company. The solution to this problem lies in a comprehensive package, which recognizes the source of the sediments and their movement dynamics. It would be necessary to retain the use of sediment traps placed at right angles to the shoreline. These sediment traps made up of rock boulders should extend for 150 m into the 2 km wide estuary at this reach of the shoreline. These protective moles would bring about differential erosion and accretion in the down drift and up drift sections of the beach respectively in accordance with the direction of the littoral drift. The semi-diurnal tidal regime of the Nigerian coast leads twice daily to current reversals, involving the ebb and flood currents. Ebb currents result in the loss of millions of cubic meter of sediments eroded from the morphologically high fossil sand bars, which are not liable to flooding. Climate change is bringing about a higher tidal range, which in turn increases the stage of the seasonal floods. Presently, in the wet season, the water level in the creeks and tidal basins in the Niger Delta is higher by over 1.5 m while the stage of rivers further upstream in the region is over 4.5 m higher than normal. Currently, only 30% of the total land area in the Niger Delta wetlands is not liable to flood at the peak of the wet season during the highest spring tide. With rising sea level actuated by climate change, this non-floodable area will be drastically reduced if a comprehensive flood protection scheme is not put in place as soon as it is practicable to do so.

The endemic problem of flooding, siltation and sedimentation in the coastal zone can only be permanently solved if the region is transformed along the pattern of the development of the Rhine Delta in the Netherlands. It may be necessary to construct polders along the margins of the tidal flats with a view to draining and reclaiming them for paddy rice and plantain cultivation on a massive scale. This will put investment capital in the hands of the people and make the development of the region sustainable. A system of canals may need to be put in place and integrated with the drainage of the region to ease transportation and improve the tourist attraction of the region. A comprehensive flood protection scheme similar to the type in the United Kingdom will have to be put in place to protect the settlements from flooding. Excess tidal flows can be impounded in large reservoirs and released for hydroelectric power generation. This will lead to large-scale resettlement of the affected local population in modern towns equipped with adequate infrastructure to make livability better than it is at present. There is also a dire need for the Government of Nigeria to set up without further delay 10 coastal tide gauges with global positioning systems (GPS) along the coastline of the country from Lagos to Calabar to monitor sea-level changes and other littoral data necessary for a proper understanding of the dynamics involved and thus be able to predict flood and erosion hazards. This will enable the population along the coast to be warned in advance of impending flood disasters and arrange for evacuation to safer grounds.

Conclusion

While it is true that climate change is controlled by natural variations, the orbit of the earth in its rotation on its axis and revolution round the sun, plate tectonics and the atmosphere-ocean exchanges of carbon dioxide and heat, the recent climate change has resulted mainly from human activities. These anthropogenic forcings of climate change include the burning of fossil fuels for power generation, deforestation, cement production and agricultural practices. These activities increase greenhouse gas emission with its concomitant greenhouse effect and increases in the temperature of the troposphere. The recent changes in climate will remain with us for a long time yet and its main impact in Sub-Saharan Africa is more extreme weather patterns and a rising sea level which will inundate about half of the present coastal belt in fifty years from now. The time is therefore ripe to begin to make contingency arrangements to understand the problem more intimately by setting up the relevant institutions and infrastructure with a strong research and development bias to troubleshoot problem areas and warn endangered population on time about impending dangers to facilitate quick evacuation. The situation that currently exists in the region in which these monitoring activities and institutions are nonexistent does not portray our communities of man as forward looking.

We must as a people, demonstrate to the rest of the world that we have the political will and the sincerity of purpose to urgently empower our Universities and Research Institutes to have functional climate stations and littoral observatories equipped with GPS tidal gauges for monitoring changes along the coastline of Sub-Saharan Africa. Afforestation schemes of the marginally arid sections of the Sahara Desert must also be urgently encouraged. The governments of Sub-Saharan African economies must of necessity discourage the decimation of forests through illegal logging and harvesting of fuel wood by substituting the use of wood with gas for domestic use. Nigeria with its vast reserves of gas can begin to harness the associated gas that is presently flared in petroleum development for domestic and industrial use in the sub-region. Research and planning globally must be directed towards adapting our food production processes towards varieties which yield optimally with declining freshwater availability. These imperatives are needful if we must effectively manage the declining resources of the sub-region in an increasingly warmer world."

References

Allen, J. R. L., 1965. Coastal geomorphology of Eastern Nigeria, Beach Ridge barrier island and vegetated tidal flats. Geologieen Mijnbbouw, 44:1–21.

Christopherson, R. W., 2002. Geosystems – An introduction to physical geography. Prentice Hill, New Jersey (Pub.).

Ebisemiju, F. S., 1988. Human impact on marine processes in the Western Niger Delta. In: Environmental Issues and Management in Nigerian Development. Sada, P. O. and F. O. Odemerho (eds). Evans Brothers. Fawora Renardet Associates, 1981. Victoria Beach Stabilization Studies. Report for Federal Ministry of Works, Nigeria.

Fubara, M. J. and Alabo Dagogo, 2007. Sustainable flood hazard mitigation in the Niger Delta. Paper presented at a National Workshop on Environmental Degradation and Sustainable Livelihoods in the Niger Delta, 26th–28th February, 2007. Rivers State University of Science and Technology, Port Harcourt.

Houghton, R. and G. Woodwell, 1989. Global Climate Change. Scientific American. April 1989, p.36.

Ibe, A. C. and E. E. Antia, 1983. Preliminary assessment of the impact of erosion along the Nigerian shoreline. NIOMR Technical Paper. No. 13, 17pp.

IPCC, 1996. Climate change 1995: The science of climate change. Contribution of Working Group 1 to the Second Assessment Report of the Inter Governmental Panel on Climate Change (Houghton, J. T., Meira Filho, L. G., Callander, B. A., Harris, N., Kattenberg A. and K. Maskell (eds.). Cambridge University Press, Cambridge, United Kingdom and New York, NY, USA. 572pp.

IPCC, 2007. Climate change 2007: The physical science basis. Executive Summary for Policy Makers, IPCC.

King, C. A. M., 1987. Ocean processes. In: Human activity and environmental processes, Gregory, K. J., Walling, D. E., John Wiley and Sons, (eds.). Chichester, 117–144.

Lawson, B. A., 1988. Coastal erosion in Nigeria. Review of its causes, effects and solutions with special reference to projects in Rivers State. Paper presented at Geomorphological Workshop of the 31st NGA Conference, April 6- 9, 1988, University of Port Harcourt, Port Harcourt.

Miller, G. T., 2004. Environmental science – Working with the earth. Thomson Brooks/Cole (Pub.). USA.

Nedeco, 1961. The waters of the Niger Delta. (Netherlands Engineering Consultants), The Hague, 1–130.

Oyegun, C. U., 1989. Beach erosion and accretion at Forcados Beach, Nigeria. Indian Geographical Journal, Vol. 64, No 2, pp. 127–140.

Oyegun, C. U., 1991a. Spatial and seasonal aspects of shoreline changes at Forcados Beach, Nigeria. Earth Surface Processes and Landforms. Vol. 16, No. 4, pp. 293–304.

Oyegun, C. U., 1991b. Land loss at Forcados South Point, Youbebe, Nigeria. WAJOST, Vol. 1., No 1, pp. 78–91.

Oyegun, C. U. and E. T. Ologunorisa, 2002. Climate change and its implications for mankind in the 21st century. JOGET, Vol. 4., No. 1, pp. 73 -84.

Oyegun, C. U., 2007. The paradox of economic and environmental development in the coastal zone of Nigeria. National Workshop on Environmental Degradation and Sustainable Livelihoods in the Niger Delta, 26th–28th February, 2007. Rivers State University of Science and Technology, Port Harcourt.

Pugh, J. C., 1954. A classification of the Nigerian coastline. Journal of the West, African Science Association. Vol. 1 (1) :1–12.

Ruddiman, W. F., 2003. The anthropogenic greenhouse era began thousands of years ago. Climate Change, 61.3: 261–293.

Ruddiman, W. F., 2005. Plows, plagues and petroleum: How humans took control of climate. Princeton University Press (Pub.).

Ruddiman, W. F., Vavrus, S. J. and J. E. Kutzbach, 2005. A test of the over due glaciations hypothesis. Quarterly Science Review, 24:11.

United Nations Environmental Programme (UNEP, 1995). Annual Report, 1995.

Usoro, E. J., 1971. Recent rates of shoreline retreat at Victoria Beach, Lagos, Nigeria. Nigerian Geographical Journal, XIV:49–58.

II. Measures and strategies for facing climate change in Sub-Sahara Africa

Emmanuel Uzoma Onweremadu
Federal University of Technology, Owerri, Nigeria

Information and communication technologies for creating awareness of climate change phenomenon in Nigeria

Abstract
Climate change is obvious and of global dimension. Impact varies with location and humans are drivers of global warming and consequent climate change. Nigeria as an African country contributes little to the atmospheric alteration mainly via agriculture, but suffers greatly from its negative effects. Information and communication technologies (ICTs) are scanty and knowledge of climate change is inadequate to enable its adaptation and mitigation. There is a variety of audience needing messages on climate change, and they require varying levels of information on the changing atmosphere for efficacy of contribution to its mitigation. Climate change communication is important as public awareness is a key to making a real difference in fighting it in Nigeria. Key political messages should focus on climate change as a threat to sustainable development. Government policies and actions should be pro-poor as a greater percentage of the populace is living below the poverty line. Such policies should originate from the opinion of indigenous people. Adaptation of communication techniques should be based on indigenous knowledge systems, a majority of which are cheap and known, while new ICTs should be tailored to conform to the existing techniques as well as socio-economic status of land users, thereby making them readily adoptable.

Introduction

Climate change is obvious and has taken a global dimension. It is now apparent that dealing with climate change is inevitable. In Nigeria, the change is tremendous resulting chiefly from burning of rangelands and croplands. In northern Nigeria nomads burn old chaffy rangelands and pastures to stimulate sprouting and re-emergence of fresh and succulent ones. In the middle and southern Nigeria with thicker vegetation of varying floristic composition, land clearing is by slash-and-burn technique. In large farms, tree pushers are used. Farmers in Nigeria practice conventional tillage instead of conservation tillage, and this involves heightened disturbance and exposure of soil aggregates. Soil disturbance increases its vulnerability to degradation forces of wind and water erosion. It is important to note that the ability to capture and store carbon in soils is a function of texture, structure, rainfall, temperature, farming system, and soil management (Lal, 2004), tillage (Onweremadu *et al.*, 2007a), cropping intensity (Ortega *et al.*, 2002) and nitrogen inputs to soils (Potter *et al.*, 1997). Tilling soils releases carbon in form of carbon dioxide, a main greenhouse gas (GHG) which accounts for about 70% of human-induced GHGs. While western countries are responsible for the largest volumes of emissions by burning fossil fuels, southern nations are contributing to global warming and climate change through agriculture. Apart from carbon dioxide, agricultural activities, especially livestock farming activity plays a significant role in producing methane, a second most important GHG. Again, dinitrogen oxide is produced by intensive agriculture, deforestation and use of chemicals such as aerosols. The solar radiation spectrum contains wavelength regions capable of causing chemical reactions in biological tissues, and much of this damaging radiation is filtered out by the atmosphere. But, it is observed that the stratospheric zones, which serve to remove a part of the spectrum known as UV-B, is decreasing globally due to increasing inputs of chlorine and bromine species into the stratosphere. These species and their sources are as follows: Chlorofluorocarbons (used as foaming agents, propellants and refrigerants), methyl chloroform and carbon tetrachloride (used as industrial solvents), halons (used as suppressants). Other trace gases that regulate the stratosphere include carbonyl sulphide (COS), carbon monoxide (CO), hydrogen (H_2) and non-methane hydrocarbons (NMHC). All these gases have natural sources but their anthropogenic emissions have now become so great that their atmospheric cycles are significantly perturbed, as evident by their continuously increasing atmospheric concentrations.

Humans are highly implicated in most drives of atmospheric change, which are crop production, animal husbandry, deforestation, biomass burning, fossil fuel burning and industrial activity. Biomass fuels such as fuel wood, crop residues and animal dung provide 85 to 90 percent of domestic energy in rural areas. These anthropogenic activities produce climatically and chemically active atmospheric trace substances, including amounts of photochemical smog from the combustion of fossil fuels (Andreae, 1991).

The effects of climate change are obvious including extinction or total loss of plant and animal species, declining biodiversity, rise in sea levels, reduced farm yields, inundations, drought, and displacement of human settlements, water shortages, increased breeding of malaria parasites, rise in water-borne diseases and increases in cerebrospinal meningitis (CSM) and heat stroke. It is expected that climate change will further accelerate the ongoing degradation processes, and in many cases leading to a complete collapse.

Throughout history, land users have always had to cope with the environment that surrounds them, gradually adapting to it. Human response then emerged from good information and knowledge of the environmental problem. Awareness on climate change varies among developed and developing countries of the world. Most developing countries are unaware of climate change while those who are observing and feeling changes lack quantitative data on it. Some developing countries with scientific data are unable to relate such to indigenous knowledge. The data become incompatible and cannot be adopted in indigenous mitigation technologies.

The total lack or insufficiency of data on climate change make developing countries such as Nigeria more vulnerable and less able to adapt to these changing atmospheric conditions. Poor and developing countries are more vulnerable to negative effects of climate change because of their locations, greater dependence on agriculture and natural resources, large variations in weather conditions and lower availability of critical resources like water, production inputs, land, capital and public services.

Little or no response to climate change due to ignorance or any other incapacitation has serious global economic consequences. Because the current world activities are knowledge-driven, it becomes very important to suggest information and communication techniques that can be readily adapt to the rural setting in Nigeria. Operational information on climate change includes information that is practical, concrete and able to mitigate existing negative effects of the change. Such information is practical if it can help in professionalizing its activities; concrete and operational if it contains real-life, tested and proven techniques and implementation methods on climate change.

Characteristics of the Nigerian rural setting

Rural landscapes in Nigeria have scattered production units and a multiplicity of small units and they are dispersed over the entire territory. The production units have great dependence on the natural environment. Structurally, these production units are little organized and more complex. Its professionals attain little formal education and are mostly associated with poor practical reading and writing skills. Language difficulties are common. Characteristically, professionals and support organizations play a central role in structuring the rural area especially in information transmission. In addition to the above, children from poorer families are less likely to go to school, which is an information and communication-enhancing institution. Such children are deprived of knowledge about their environment.

The above specificities influence information and communication in rural areas. Material and languages (extension) are needed for information dissemination. This becomes demanding as the rural world is often on the sidelines of information channels, particularly new information technology. In a rural setting, the role of professional and support organizations is central and built around fostering frameworks of dialogue, networking and the capacity to collect and disseminate information on climate change.

Information needs for policies on climatic change

Information is the bedrock of policy pillars, and indeed the main instrument at every stage and level of policy making. All stakeholders in climate change have access to information relevant and appropriate to mitigating the effects of the changing atmosphere. Information and communication technology (ICTs) are expected to capture, store, process, share, display, protect and manage information on climate change.

There is need for stakeholders in climate change to access information for the purpose of identifying emerging issues and problems, advocating and sensitizing wider audience on environmental issues, participating and contributing to policy design and implementation, complying with polices, legislations and regulations, mediating and negotiating where applicable and innovating approaches, institutions and technologies. Insufficient, contradictory and outdated information result in poor assessment of climate change and social-economic factors affecting the change. Information on the optimum productive capacity of natural resources and environmental degradation as influenced by climate change, its implications and causes is essential in order to make appropriate decisions on resource conservation. Since many issues on global warming and climate change are anthropogenically-oriented, natural resources management policy-making require information on economic and socio-cultured conditions in which natural resources are used.

It is important to note that all stakeholders, not only technical institutions are users, providers and managers of information (Brahimi, 2000), and need to be aware of the various information sources (Katerere, 2000) located both within and outside the public sector including local communities, ministries, research institutions, farmers support services, civil society organizations, private companies, consultants and international organizations.

Access to information on climate change can be promoted through awareness building and debates, networking and consultations, disseminations of research findings, public education and sensitization. Access to sufficient and relevant information on global warming and climate change by land users is a basic precondition to people to participate in and contribute to policy decision-making.

Information generators and providers may face the problem on how to reach policy-makers and make them use appropriate information when taking decisions on climate change. Generating adaptable and adoptable information and communication technologies may face the problem of their use by decision makers. Generally, decision makers may not use large volumes of statistical tables and reports written in complex technical language as given by information experts and providers on climate change. Great volumes of data are obsolete, unreliable and lacking in historical sequence, and with a display of disparity hence not exploitable. Style, language and dissemination instruments need to be adapted to the characteristics of those who are possible users of the information on atmospheric changes. Furthermore, cost of information adds another dimension to management of abnormal climate variations. This becomes an issue in mitigating climate change in a country by poor rural dwellers who contribute significantly to the change via rural-based agricultural and non-agricultural activities aimed at sustaining life. It is often argued that information should be paid for by rural stakeholders in environmental management.

Certainly, new information technologies on climate change can spark off positive changes on adaptation measures, policies and strategies. It is known that the choice of adaptation measures is constrained by lack of knowledge of how to implement them coupled with other factors such as expense, countervailing beliefs and cultural practices. But provision of technical information, advice or guidance as well as weather and seasonal climate forecasts and warnings to land users at risk from climate change puts them on safe landing grounds.

New ICTs generate a larger volume of climate information available which can also be juxtaposed if they are of different natures. Other advantages of ICTs include the decentralization of information sources, a multiplicity of producers combined with less distance between information users and faster communication and greater ease in network operations.

Rapid diffusion of ICTs is a *sine qua non* for productivity and growth in developing countries. Relative lack of literacy and numeracy typically characterize the Nigerian poor as does the lack of access to accurate market prices, technical and other information relevant to the profitability of their agribusiness decision and their reintegration with markets. Illiteracy and lack of education breed social and cultural isolation and the poor who are often in remote places are further handicapped by limited availability of public information. Adequate and accurate information on climate change is critical if the poor are to make progress in livelihood decisions that can enable them escape poverty. Educating land users is critical to conservation of natural resources. Empirically, education has been shown to have significant positive impacts on agricultural productivity, engaging in off-farm self-employment, getting a job, overall income, accessing credit, size of social network, political participation, using government services, and own health and child health outcomes (Ahmed *et al.*, 2007).

Target audience in climatic change awareness

There is a causal nexus between natural resources, population increase, poverty and environmental degradation. It follows that the most deprived persons, most who live below poverty line, do contribute significantly and negatively to climate change in their bid to survive. A great many of this category of humans live in Sub-Saharan Africa, especially in the most remote areas with higher incidence of poverty.

Information services on climate change aim to build an offer of response to identified needs. Useful information on climate change is much more desirable than obtaining large volumes of data that cannot be readily accessed and utilized.

Audiences in climate change awareness include small individual farmers, farmers' organizations, project implementers, small and medium enterprises, agricultural workers union, support and advice organizations, and teacher and students. These and other categories of humans need information to enable them play active roles in the mitigation of climate change. This may range from policy, through technical commercial and markets, legal and fiscal instruments of financing and management of information. Information is matched to each target audience. Matching audience to information needs involves knowing how the information is obtained, types of information demanded and the level of information available for use. Identification of the audience is a good step towards solving problem of climate change awareness programs.

Climate change communication (CCC) for Nigeria public awareness is the key to making a real difference in fighting climate change in Nigeria. Every government, no matter its ideological stripe, recognizes some role of communication in national development. A developing nation like Nigeria needs the media for the evolution of an environmentally-friendly personality. Applying traditional knowledge and cultural practices that help reduce greenhouse gases emission has proved to be an effective tool to communicating on climate change.

A successful factor in climate change communication is to localize a global problem. Communication for climate change needs a personal level. Climate change should be made a home issue since humans contribute in a variety of forms. For instance, every home uses different forms of energy, hence the need for establishing energy efficiency and renewable energy as a routine best practice (Dilling and Farhar, 2007). A second factor to effective climate change communication is making it readily understandable by all irrespective of literacy level. Messages on climate change rely on clarity and simplicity in order to attain efficacy. Fewer words in message make greater impact. Climate change communication strategies need to create a sense of team work among land users or stakeholders. Unity is strength, thus collective power and effort of the family, neighbors, colleagues, relations and friends can certainly have a larger impact. Nigerians like most other Sub-Saharan Africans are traditionally communalistic rather than individualistic, stressing the relevance of community effort in climate change communication strategies.

In another dimension, the audience should be made to relate the concept to something with which stakeholders are familiar. Incorporating culture and social values creates a sense of place, relating climate change to family, surroundings and immediate catchment areas.

Government leadership factor is important for effective climate change communication. Leaders are expected to show example. The success of government campaign comes when it demonstrates necessity for taking action. Government should lubricate the leverage approach where a large awareness campaign is drawn to reach out to a wide audience. It is government that can fund the leverage approach, since it entails a large amount of resources and effort.

Understanding the need for these attributes in climate change communication goes a long way to changing the attitude as many still do business as usual. Ineffective communication strategies constitute a major barrier to motivating and empowering the public to take action on the changing atmosphere conditions.

Political awareness

Africa has the least GHG emissions but is hit hardest by climate change. As climate change is compounding existing risks and vulnerabilities in Sub-Saharan Africa, it calls for political leadership to make commitment to the challenge. Nigeria must take proactive approaches in negotiating for a global GHG target that prevents dangerous climate change, and should be ready at all times to join the international community in conferences and summits on this hotspot. This is in addition to mainstreaming adaptation into development.

Key political messages and action points should focus on climate change as a major threat to sustainable growth and development in Nigeria, and the achievement of the Millennium Development Goals. This becomes necessary as Nigeria is vulnerable to climate change effects, which include worsening food security, increased incidence of flooding and drought, spread of diseases, increased risk of conflicts over scarce land and water resources.

Climate observations and data management play key roles in disaster management and climate adaptation. Awareness obtained within the National Meteorological and Hydrological Services and the climate community on general development as well as specific information needs such as agricultural extension workers and rural communities have proved efficacious in some non-Nigerian locations. Nigeria can lead the bubble in West Africa or Sub-Saharan Africa, where it joins hands with other neighboring countries to meet targets jointly via aggregation of their total emissions. Such leadership in itself is exemplary and thus creates awareness.

At the national scene, Nigeria should establish a National Coordination Committee on climate change. A series of policies and measures to address the change should be part of the national development strategy, making positive contributions to the mitigation and adaptation to climate change. Trends of climate change should be observed and monitored over major agro-ecological zones of Nigeria.

Climatic characteristics such as rainfall, temperature, wind pattern, isolation, pressure and relative humidity should be observed in major zones. In addition, the sea level along the Atlantic coast should be measured. Records on current GHG emissions are important for policy decisions especially as it affects CO_2, CH_4 and N_2O levels.

Nigeria should use results of biophysical surveys to restructure the economy, promote technological advancement and improving energy efficiency. There should be optimization of energy mix by developing low-carbon and renewable energy. High grade and clean energy can be improved by strengthening the development and utilization of hydropower; and by supporting the development and utilization of new and renewable energy like biomass, solar, geothermal and wind energy.

Launching of a nationwide tree planting and afforestation campaign, enhancing ecology restoration and protection for climate change is currently demanding in Nigeria. Plant species with high mitigation capacity should be multiplied through nursery practice and shared among farms. However, Government should involve farmers actively in multiplication programs for easy adoption.

Government of Nigeria should strengthen laws and regulations, policies and measures relevant to addressing climate change. This will help in accelerating the building of a national resource-conserving and environmentally-friendly society. Laws and regulations on medium- and long-term energy development, renewable energy, price management, differentiated pricing, special fund, etc., should be put in place. Climate change research is a major priority. Also, great importance should be attached to capacity building. Support should be given to scientific studies and researches on climate change. Studies should focus on impact and counter measures of global climate change, forecasting, and global climate change and environmental policies. Fundamental research should be on carbon cycling in Nigeria's terrestrial and aquatic ecosystems, ocean level change, trends and impacts, socio-economy of climate change, etc.

Nigeria should strengthen education, training and public awareness on climate change. The Ecological Fund should be structured to promote awareness on climate change. All forms of education at all levels should be encouraged to enhance public awareness on climate change especially as it affects anthropogenic activities. The awareness must recognizes harsh and excruciating Nigerian climate conditions, vulnerability of Nigerian ecosystems, extraction of solid minerals and exploration and exploitation of crude oil, increasing population, low level economic development, high level corruption and failure of governance at all levels of her governments. National awareness programs should rely on information and other lower levels of governance. In rural areas, climate field schools should be encouraged by government for improving response farming to climate change. Farmers have always responded to variability in climate especially rainfall by adapting whenever such changes occur.

Adoption and adaptation of indigenous information and communication technologies in Nigeria

Indigenous knowledge is the local knowledge and information base for a society, which facilitates communication and decision making. It is clear that cross-cultural interactions have influenced the Nigerian indigenous knowledge systems (IKS) since IKS is dynamic and affected by internal creations and experimentation as well as contact with external system (Asiabaka, 1994). Yet, IKSs are comprehensive and conclusive (Ellen and Harris, 1996) and constitute the social capital of the poor, the main asset to invest in the struggle for survival (Asiabaka, 2009). Some IKSs are not efficacious (Onweremadu *et al.*, 2007) especially in land use techniques (Onweremadu, 2008), but they should be adapted in the light of the current needs.

In information and communication, a great number of the technologies are still relevant and adequate in transmitting messages bothering on climate change. Literacy level in Nigeria is still low, and this justifies the use of indigenous communication instruments. The most significant information gathering exercise in Nigerian IKS is oral traditional (verbal communication). It constitutes a major resource in sending messages on climate change via slogans, music ceremonially or spiritually derived language, poetry, narratives or tales as well as commentaries. While narratives are historical, instructive, artistic and personal, commentaries may be legal or non-legal. No longer is the library a depository of books, but its ambit has extended to teaching and learning strategies such as discussions, debates, symposia, interviews and tutorials. The participatory atmosphere generated by such activities is compatible with the spirit and content of IKS. The oral means of creating awareness on climate change should be researched on in the context of knowledge accumulation and consolidation, problem solving and community development.

Non-verbal means of sending climate change messages abound in Nigeria IKS. The universal purpose of communication is to stimulate reaction from the person to whom the message is addressed and Nigerian non-verbal instruments are replete with this attribute. In Igbo land, southeastern Nigeria, a number of instruments are developed for this purpose, and they include Ekwe (a wooden gong), Ikoro (big tree trunk with hollow), Ikperikpe (talking drum), Oja (wooden flute), Ogene (metal gong), Opi (horn), Kurutu (canon), Egbe (gun), Omu (palm fronds), etc. These can be improved via research for adoption and adaptation to meet communication challenges of climate change.

Traditional languages in Nigeria should be visited as media for sending messages on climate change. Language is a medium for scientific creativity and innovation, and indeed a vehicle for intellectual activity. In western culture, growth in science and technology has depended tremendously on the improvement in scientific communication and intercourse between scientists and laymen, many of whom have made significant scientific discoveries.

Furthermore, current information and communication technology tools should be brought within the reach of smallholder agriculturists and the likes. In other words, the ICTs should be pro-poor as about 70% of Nigerians are living below poverty line, and depend on natural resources for survival or face extinction in the biosphere. Low income rural residents can adopt some ICTs with considerable success. Mobile phones are adoptable in Nigeria given its relatively inexpensive cost. It can give accurate and real-time information on climate change. Internet technology can be useful in urban and peri-urban centers of Nigeria. Telecenters and market information systems (MISs) can be useful in sending messages on climate change in the Nigerian rural landscape. Information service on climate change can be located at market centers, churches, mosques, and timed to target popular festivals and ceremonies in Nigeria. Use of charts, radio and television will be a good strategy for creating awareness on the changing atmosphere. Policies on ICTs should be tailored to create congenial climate for high rates of investment by private enterprise, wire small scale land users into connectivity while archiving indigenous knowledge related to farm extension, converting it into local vernacular and populate an email network with farmers interested in receiving farm extension online. Again, government policies should favor provision of smallholder farmers with leading-edge computer hardware, enabling applications that improve productivity while promoting farmer-friendly web content. Wiring schools into the internet, exposing children to computerization and making computer laboratories in such schools into community learning hubs will be helpful. On climate change issues, children in such communities can learn during regular hours, while their parents come into class at after-hours as continuing education students. In all, both formal and informal education on climate change will go a long way in promoting capacity building to the recipients. In conclusion, all these information and communication strategies should be built on the existence of local initiatives from individuals and institutions. This will inspire and change current attitudes of land users towards environmental conservation.

References

Ahmed, A. U., Hill, R. V., Smith, L. C., Wiesmann D. M. and T. Frankenberger, 2007. The world's most deprived: characteristics and causes of extreme poverty and hunger. International Food Policy Research Institute, Washington D.C., 2020. Discussion paper, 43, 130 pp.

Andreae, M. O., 1991. Biomass burning in the tropics. Impact on environmental quality and global climate. In: Davis K. and M. S. Bernstam (eds.). Resources, environment and population: Present knowledge, future options. Pp. 268–91. New York: Oxford University Press.

Asiabaka, C. C., 1994. Indigenous system in cassava plant protection. Proceedings of the expert meeting on indigenous knowledge system for sustainable agriculture. National Root Crop Research Institute, Umudike. Pp. 1–4.

Asiabaka, C. C., 2009. Indigenous knowledge systems (IKS) in organic agriculture. In: Obiefuna, J. C., Udedibie, A. B. I., Osuji, G. E., Ofoh, M. C., Osuigwe, D. I., Onweremadu, E. U. and E. B. Etuk (eds.). Organic agriculture for health, wealth and environmental conservation. Proceeding of the 5th National Conference on Organic Agriculture Project in Tertiary Institution in Nigeria at Federal University of Technology Owerri, Nigeria, 15–19th November, 2009. Pp. 165–172.

Brahimi, Y., 2000. Monitoring, evaluating and exchanging information on desertification: The implementation on the CCD in Africa. In: Proceedings of a CTA workshop on information support for natural resource management policy, Wageningen, The Netherlands, 26–29 January, 1999. Pp. 99–110.

Dilling, L. and B. Farhar, 2007. Making it easy: Establishing energy efficiency and renewable energy as routine best practice. In: Moser, S. C. and L. Dilling (eds.). Creating a climate change: Communicating climate change and facilitating social change. Cambridge University Press, Cambridge, UK. 549 pp.

Ellen, R. H. Harris, 1996. Concept of indigenous environmental knowledge in scientific and development studies iterative: A critical assessment, a draft paper, East-west Environmental Linkage Network Workshop, Canterbury.

Katerere, Y., 2000. Who makes policy and how? The case of community-based natural resources management in Southern Africa. In: Proceeding of a CTA workshop on information support for natural resources management policy. Wageningen, the Netherlands, 26–29 January, 1999. Pp. 31–51.

Lal, R. 2004. Soil carbon sequestration impact on global climate change and food security. Science 304:1623–1627.

Olowo, V. I. O., Enikuomehin, O. A., Adeyem, Y. A., Adebimpe, O. A., Adeniregun, O. O. and M. Abiala, 2009. Climate variability and sunflower yield in the transition zone of Southwest Nigeria. In: Obiefuna, J. C., Udedibie, A. B. I., Osuji, G. E., Ofoh, M. C., Osuigwe, D. I., Onweremadu, E. U. and E. B. Etuk (eds.). Organic agriculture for health, wealth and environmental conservation. Proceedings of the 5th National Conference of Organic Agriculture Project in Tertiary Institutions in Nigeria at Owerri, Nigeria, 15–19 November, 2009. Pp. 14–20.

Onweremadu, E. U, Asiabaka, C. C., Adesope, O. M. and N. S. Oguzor, 2007. Application of indigenous knowledge on land use activities among farmers in central southeastern Nigeria. Online J. Earth Sci; 1(1): 47–50.

Onweremadu, E. U., Onyia, V. N. and M. A. N. Anikwe, 2007. Carbon and nitrogen distribution in water stable aggregates under two tillage techniques fluvisols of Owerri area, Southeastern Nigeria, Soil Till Res, 97: 195–206.

Onweremadu, E. U., 2008. Evaluating soil structure and hydraulic conductivity by land use in Nigeria. Soil Survey. Horizon, 49:6–11.

Ortega, R. A., Peterson, G. A. and D. G. Westfall, 2002. Residue accumulation and changes in soil organic matter as affected by cropping intensity in no till dry land agro ecosystem. Agron. J. 94: 944–954.

Potter, K. N., Jones, O. R., Torbert, H. A. and P. W. Unger, 1997. Crop rotation and tillage effects or organic carbon sequestration in the semiarid Southern Great Plains. Soil Science, 192: 140–147.

Cyril Ifeanyi Duruigbo
Federal University of Technology, Owerri, Nigeria

Strategies adopted in preserving plant and animal species endangered by climate change in Sub-Saharan Africa

Abstract
The rate of biodiversity and ecological degradation in Sub-Saharan Africa is alarming due to climate change impacts. The application of farmers' indigenous knowledge in cover cropping, agro forestry systems, use of botanic gardens, adaptation measures, regenerative agricultural practices, establishment of game and forest reserves, breeding crop and animal varieties that adapt to climate change, setting up of gene banks and building a strong environmental protection and information policy by government in Sub-Saharan Africa will go a long way to combat the impacts of climate change and preserve endangered biodiversity in the ecosystem.

Introduction

The concept of biodiversity refers to an index of species richness or the number of distinct plants and animals in a given environment (Srivastava *et al.*, 1996). The continued productivity of existing crops and livestock hinges on a large part on harnessing the genetic variation within each species. The conservation and management of biodiversity is particularly critical in lower latitudes, since they contain more species than are typically found in temperate areas. Tropical rain forests cover about seven percent of the earth's surface but more than half of the world's biota (Wilson, 1988). Tropical forests in Sub-Saharan Africa contain wild populations of hundreds of crops such as mango, coffee, rubber, guava as well as candidates for new crops and livestock species. Habitats that have been modified for crop and livestock production are known as agro-biodiversity. The effect of climate change are obvious including extinction or total loss of plants and animal species, declining biodiversity, rise in sea levels, reduced farm yield, inundations, droughts, displacement of human settlements, water shortages, increased breeding of malaria parasites and disruption of the eco-system. Significant climate change is expected over the 21st century and it will affect eco-systems and access to natural resources such as fertile land and water (IPCC, 2007). Regions with low adaptive capacity due to poverty, lack of infrastructural services and good governance such as Sub-Saharan Africa will be severely affected.

Sub-Saharan Africa is housing about 20% of all known species of plants, mammals and birds. This rich biodiversity is currently being degraded due to climate change impacts (Boko, 2007). Mountainous regions are projected to suffer from temperature increase most prominently demonstrated by the melting of the glacier on Mt. Kilimanjaro (Boko, 2007). Under climate and land-use changes, habitats are projected to shift spatially or disappear completely forcing species to migrate (McLean *et al.*, 2005). About 5000 Africa plant species have been negatively impacted due to climate change. In South Africa, Fynbos and succulent karoo biomes losses are estimated at 51–61%; critically endangered taxa, for an example, the predecease is lost with 2% of the 227 taxa becoming extinct (Boko *et al.*, 2007). As with any ecosystem, the functioning and sustainability of agricultural systems depend greatly on biological diversity, thus efforts made by farmers, environmentalists and various stake-holders for healthy and stable environment is highly commendable, especially in the face of climate induced ecological degradation. This paper highlights the various strategies adopted by farmers and government in Sub-Saharan Africa to preserve biodiversity a risk of degradation and extinction.

Strategies adopted by farmers and government in preserving species endangered by climate change

Indigenous knowledge approach

The history of natural resource management in Sub-Saharan Africa started some two million years ago as human beings tireless pursuit to provide basic food and cultural needs through hunting and agriculture. Over time, humans developed their unique capacity to think, adapt to changes, foresee, accumulate and pass on intellectual wealth to subsequent generations (Ewert *et al.*, 2004). Farmer's traditional ecological knowledge is historical, cumulative and dynamic, building upon their earlier experiences from their ancestors and adapting to the socio-economic changes of the present (Berkes, 1999; Doubleday, 1993) African farmers have long devised effective measures to manage natural resources and mitigate land biodiversity degradation. These coping strategies are as follows.

Planting of cover crops

Farmers normally include cover crops such as sweet potato *(Ipomoea batatas)*, Egusi melon (*Colocynthis citrullus* and *Mucuna pruriens*) to help in soil moisture conservation, weed suppression and provide good micro-climate for the component crops. Cover cropping increase the activities of soil micro-fauna, macro-fauna as well as micro-flora, respectively. This increased biological activity helps in stabilizing the ecological system and thus increase the survivability of species that would have been endangered due to high temperatures.

Agro-forestry systems approach

African farmers usually deliberately retain certain tree and shrub species in their crop production system, because they recognize that these tree or shrub species have long term beneficial effects. This practice by farmers will enhance soil fertility, prevent soil erosion, control weeds, store more carbon, and increase crop yield. Bene *et al.* (1977) defined agro-forestry system as a "sustainable management system for land that increases overall production, combines agricultural crops and forest plants and animal simultaneously or sequentially and applied management practices that are compatible with the cultural pattern of the local people." Agro-forestry eco-system is extremely dynamic with available resources capable of coping and containing global climate change in the long run.

Establishment of botanic and zoological gardens

Farmers deliberately collect high medicinal value plant species that are facing extinction and cultivate them in specially managed botanic gardens in their homesteads. Such plant species include: *Azardiracta indica* (Neem Plant), which is valued for use in the cure of malaria and other ailments, *Garcinia cola* (bitter kola), *Cola nitida*, *Cola acuminata* (kolanuts), and *Butyrospermum parki* are among the agro-forestry multi-purpose plants supplying cooking oil, and serving as skin cream for burns and wounds. These plant species indicate great potentials for agro-industries, energy and biotechnology hence their preservation in botanic gardens will ensure their continued existence in the face of climate change impacts. Farmers also hunt and domesticate wild species of animals that are facing extinction due to harsh weather. These compound zoological gardens help to preserve the endangered species that would have been lost in the wild. In some African households, such endangered species of animals such as birds are caught in the wild and kept as pets in the homes.

Adaptation to environmental stress (Climate change)

Adaptation is an automatic or planned response to change that minimizes the adverse effects and maximizes any benefits. It is one of the two possible means of coping with human induced climate change and sea level rise. Farmers' response to this direction is by selecting local crop and livestock species that adapt to the unfavorable climate. The building of dams to control flooding and water storage devices such as ponds to supply water in times of drought has also been in practice by African farmers for a long time. Farmers also change the crop management practices such as increasing irrigation and altering the planting dates to further adapt to the changing weather situation. In many situations farmers already adapt to variations in seasons on a year-to-year basis by planting later in the season. Climate impacts and the adaptive capacity of societies determine their vulnerability to climate change.

Application of regenerative agricultural practices

Sustainable agriculture as a local practice and research priority has been used by farmers in response to wide-spread recognition of the need to balance food production with a healthy environment that will sustain biodiversity (NRC, 1989). The sustainability of employing mechanized, high-input and specialized agriculture is being questioned due to its negative impact on the environment. Farmers in their wisdom adopted agro ecology, which involves the use of ecological principles for the design and management of natural resources and conserving agricultural systems. These systems rely practically on farmers indigenous farming knowledge and selected modern technologies to manage diversity and incorporate biological principles and resources into farming systems to boost food production, reduce negative impacts of climate change and preserve endangered species of plants and animals (Pretty, 1995).

In Sub-Saharan Africa, evidence shows that regenerative and resources conserving technology and practices also adopted by farmers have both environmental and economic benefits to farmers, communities and biodiversity. In these areas farming communities adapting regenerative technologies have substantially improved agricultural yields often with a few or no external input (UNDP, 1992; Shah, 1994; Pretty, 1995). The ability of the soil to capture and store carbon is a function of texture, structure, rainfall, temperature, farming system and soil management (Lal, 2004), tillage (Del Grosso *et al.*, 2008), cropping intensity (Ortega *et al.*, 2002), and nitrogen inputs to soils. Tilling soils releases carbon in form of carbon-dioxide, which is a main greenhouse gas accounting for about 70% of human-induced GHGs. While western countries are responsible for the largest volumes of emissions by burning fossil fuels, southern nations are contributing to global warming and climate change through agriculture. This is why the adoption of good agricultural practices by African countries is inevitable firstly to boost food security and secondly to conserve and preserve the environment for the continued existence of biodiversity. The adoption of this ecologically-beneficial type of agriculture can be facilitated through favorable environmental policies, motivation and participation by farmers. The adoption of conservation tillage which minimizes soil disturbance with adequate soil residue cover contributes to soil organic matter in the arid and semi-arid areas of Sub-Saharan Africa. Conservation tillage reduces erosion and carbon losses that are critical in climate change. It is estimated that with careful soil management, the potential carbon sequestration under conservation tillage per year could reach 0.05–0.1 $mgha^{-1}$ for rain-fed crops and range-land (Rosenberg *et al.*, 1999). Conservation tillage has the potential to mitigate greenhouse gas effect, improve soil structure, increase soil water infiltration, soil water holding capacity and enhanced biodiversity activities. In West and Central Africa the natural recycling of animal droppings through its disposal on cultivated land for crop production is a good soil fertility management strategy, while reducing dependence on mineral fertilizer (Ogunwole, 2004).

Establishment of forest and game reserves by governments

Most governments in Sub-Saharan Africa help in the protection of species endangered by climate change by establishment of forest reserves and zoological gardens, while important tree crop and animal species are protected through experienced forestry and livestock professionals who cater for the needs of the trees and wild animal species. The tree and wild animal species provide tourist attraction centers and thus bring revenue to the government. For instance, in South Africa, Mkhuze and Phinda game reserves were established to protect South Africa's valuable ecosystems from going into extinction. In Kenya, the Sweetwater Black Rhino Reserve, a 100 square kilometer reserve was set up to house black Kenyan Rhinos. In Nigeria, the Yankari game reserve and the Abuja zoo are efforts of government to safeguard the existence of endangered species in the eco-system.

Protecting biodiversity through the establishment of gene banks

Government should support the establishment of gene banks that will contain various crop germplasm. These germplasm contains landraces or wild species of crops being threatened by climate change in the wild. There is need for plant breeders, geneticists and other scientists to collect and conserve crop varieties, landraces and wild relatives that are currently facing extinction due to climate change. Despite the fact that globally there are an estimated six million crop specimens contained in more than 1,300 gene banks, crop diversity has been vanishing from fields before, having been collected or dying in sub-standard gene banks due to chronic under-funding, especially in Sub-Saharan Africa (ACIAR, 2009). These biological resources contain a huge amount of un-tapped and endangered bio-diversity of inestimable value in helping agriculture cope with climate change and population growth. Adequate funding and recruitment of experienced scientists for the breeding of crop and animal species that adapt to climate change is the way to go. Plant breeders, geneticists and agricultural scientists should tactically induce traits in crop and livestock species that will resist the impacts of climate change stresses such as high temperature, drought, flooding and chilling injury.

Building a strong environmental protection and information policy

The Inter-governmental Panel in Climate Change (IPCC) was set-up to provide foresight in relation to the possible human-induced impacts on climate change with a view to assisting governments in Africa and elsewhere to formulate wiser environmental protection policy options and decisions as well as provide policy-relevant scientific advice. This followed the call by the UN General Assembly for a United Nations Framework Convention on Climate Change (UNFCC) in 1990, which was finally adopted in New York in May 1992 and ratified by 193 countries in 2007. The objective of the UNFCC includes "the stabilization of greenhouse gas concentration in the atmosphere at levels that would prevent dangerous anthropogenic interference with climate systems. Such a level should be achieved within a time frame sufficient to allow eco-systems to adapt naturally to climate change, ensure food security and sustainable economic development. Sub-Saharan African countries, which are most vulnerable to climate change impacts, should be in the forefront of articulating an environmentally-friendly policy on climate change in response to the 1997 Kyoto protocol where various parties to the convention agreed to start the process of reducing greenhouse gases to an average of 5.2% between 2008–2012.

Time is ripe for various governments in Sub-Saharan Africa to confront the challenge posed by climate change by implementing strict environmental policies that will boost food security, sustain the environment and preserve biodiversity. A series of policy measures aimed at informing, transforming and tackling the impact of climate change must be part of Sub-Saharan Africa's government development strategy. Such policy will involve mitigation and adaptation measures required to curb the impact of climate change.

African leaders should embark on policy measures that encourage tree planting and discourage deforestation, reducing burning of fossil-fuel and avoid unsustainable agricultural practices that cause environmental and biodiversity degradation. The policy option must be designed to include the setting up of Geographic Information Systems (GIS) and Early Warning Systems (EWS), mounted at strategic locations to track down climate change related data for use in adaptation, mitigation and preservation measures in the eco-system. It is also very crucial to institute a legal framework to tackle irresponsible and bad practices that degrade the environment and biodiversity. The policy option must involve the media and internet to boost climate change awareness to the public. There should also be a deliberate policy to preserve plants and animal species under extinction by establishing a special commission to cater for biodiversity. The commission shall oversea the collection, storage and preservation of endangered plant and animal species under threat of climate change. Finally, good governance, education and collective action are required from all stakeholders to protect and preserve endangered species and habitats in the eco-system.

References

ACIAR (Australian Centre for International Agricultural Research), 2009. Millennium Development Goals. In: Partners in Research for Development. ISSN-1031-1009.

Bene, J. G., Beall, H. W. and A. Cote, 1977. Tree, Food and People. Land management in the tropics. IDRC publication 0842.

Berkes, F., 1999. Sacred Ecology: Traditional Ecological Knowledge and Resource Management. Taylor, A. P. and J. A. Francis (eds.), Philadelphia. In: "The Way of Development," Blaser, Mario, Harvey, A., and Feit and Glenn Mcrae. ISBN, 842771922.

Boko, M., 2007. Africa. In: Parry, M. L. *et al.* (eds.), Climate Change, 2007. Impacts, adaptation and vulnerability. Contribution of working group II to the fourth assessment report of inter-governmental panel on climate charge. Cambridge, U. K., University Press.

Del-Grosso, S. J., Haworson A. D. and W. J. Parton, 2008. Testing DAYCENT model simulations of corn yield and nitrous oxide emissions in irrigated tillage system in Colorado. Journal of Environmental Quality. 37:1363–1389.

Doubleday J., 1993. "Finding common ground, natural law and collective wisdom." In: J. Inglis (ed.), Traditional Ecological Knowledge, Concepts and

Cases. Ottawa. International Programme. Traditional Ecological Knowledge and International Development Research Centre. Pp. 41–53.

Ewert, A.W., Baker, D. C. and G. C. Bissix, 2004. Integrated Resources and Environmental Management the Human Dimension. CABI publishing Wallingford. 290 pp.

IPCC (Inter-governmental Panel on Climate Change), 2007. Synthesis Report, Geneva.

Lal, R. 2004. Soil carbon sequestration, impact on global climate change and food security. Science 304: 1623–1627.

McClean C. J., 2005. African plant diversity and climate change. In: Annals of the Missouri Botanical Garden, 92 (2), 139–152.

NRC (National Research Council), 1989. Alternative Agriculture. Washington D. C., National Academy Press.

Ogunwole J. O., 2004. Effects of fertilizer and time of ridge tie on yield and fibre quality of late-sown cotton in the dry Savannah zone of Nigeria. Journal of Sustainable Agriculture, 24 (3): 97–107.

Ortega, R. A., Petterson, G. A. and D. G. Westfall, 2002. Residue accumulation and changes in soil organic matter as affected by cropping intensity in no-till dry-land agro-ecosystem. Agronomy Journal, Vol. 94, pp. 944–954.

Pretty, J. N., 1995. Regenerating Agriculture: Policies and practices for sustainability and self-reliance. London. Earthscan Publications, Ltd.

Rosenberg, N. R., Izaurralde R. C. and E. I. Malone, 1999. Introduction to proceedings on carbon sequestration in soils: Science, Monitoring and Beyond, Proceedings of the St. Michaels Workshop. Pp. vii–xii.

Shah, P., 1994. Village managed extension systems in India: "Implications for policy and practice." In: Jan Scoones and John Thompson (eds.). Beyond Farmer First. London, Intermediate Technology Publication.

Srivastava, J., Nigel, J., Smith, H. and Douglas Fomo, 1996. Biodiversity and Agriculture: Implications for conservation and development. Technical Paper, 321. Washington, D. C., World Bank.

UNDP (United Nations Development Programme), 1992. The benefits of diversity. An incentive towards sustainable agriculture. New York.

Wilson E. O., 1988. "The Current State of Biological Diversity." In: Wilson, E. O. and F. M. Peter (eds.). Biodiversity. Pp. 3–18. Washington, D. C., National Academy Press.

Cyprian Osinde, Bateganya Najib and Gastone Habarugaba
Makerere University, Uganda

Global climate change in Sub-Saharan Africa: The situation in Karamoja sub-region, Uganda

Abstract
This document outlines the key elements of global environmental problems, especially those arising from climate change in the semi-arid conditions of Karamoja region in Uganda. Global Environmental Change Research encompasses the interlinked issues of social, economic, political and technological change; their consequences on the land surface and its water systems in the oceans and in the atmosphere; the resultant changes in the climate; and the impact of all the above on biodiversity and human well being. Africa is particularly vulnerable to global change, partly due to its location, and also because of limited capacity to adapt to such unpredictable natural occurrences. Africa has, however, a significant capacity to do global climate change research, but is not reaching its full and necessary potentials in this arena.

Strategies aimed at solving climate change related problems involve capacity building in global climate change research at individual, institutional and policy making levels. It is proposed that this can best be achieved through large, sustained, integrative research campaigns, initiated and based in Africa, but consistent with global objectives and carried out through international partnerships.

Introduction

"Global climate change" refers to changes that are altering the contemporary earth at an unprecedented and accelerating rate. It includes the growing demands of the human population for food, fiber and water that have led to a dramatic transformation of the land surface, from quasi-natural cover to cultivated lands. The carbon dioxide liberated from this source has combined with emissions from the burning of fossil fuels to change the composition and radiant energy balance of the atmosphere, leading to a warming of the earth's surface, changes in the distribution and quantity of rainfall, and a rise in the mean sea level (IPCC, 2001). Waste products of human activities are leading to an over-abundance of nutrients in freshwater and coastal systems, accumulated toxins on land and deteriorating air quality (MEA, 2005). Unsustainable harvest pressures and loss of habitat for wild species has precipitated a loss of biodiversity; at the same time a small number of weedy species are increasing their dominance. These changes have implications for humans in various ways including constraining livelihood resources in poorer communities and by impacting tourism and the range of other goods and services that ecosystems provide (MEA, 2005). Human societies are themselves changing from rural to urban and from locally-focused to globally-connected. This document outlines key elements and problems, as well as the strategic measures aimed at mitigating the myriad of problems arising from global climate change in the Karamoja region of Uganda.

Research methodology

A team of researchers focused on selected subsets of research topics arrived at by a consultative, but ultimately subjective, search for a limited set of coherent, highly integrative themes that represented the intersection of African relevance, opportunity and capability. The aim was to design a few, big, long-lasting regional and sub-regional programs. It was agreed that no more than approximately six 'big' questions or research themes should be tackled by our research team so that long lasting and sustainable research and capacity can be generated. These topics should take note of the global climate change and Earth System Science program strategies and exploit African comparative advantages within them. It should also encourage inter-disciplinary and multi-institutional projects and programs, where appropriate research agendas should be framed in an inter-disciplinary manner to capture the multiple causality of African environmental change.

The problems

Africa is characterized by its diversity of peoples and natural environments. The continent and its adjacent islands occupy a total land area of 30.4 million km^2, slightly more than 20% of the world's landmass. It had a population of over 880 million in 2005, with a growth rate of 2–4%, twice the global mean. This growing population will exacerbate existing problems of food security and the provision of safe water, education and health services. It adds ecological stresses to the glaring economic strains evident in Africa. The key observations due to climate change facing Africa include:

Hunger, Poverty and Disease: Forty percent of the population in Sub-Saharan Africa is undernourished (Benson, 2004). Most of the countries currently facing food emergencies are in Africa. Hunger is not a sporadic episode in Africa: nearly 200 million of its people are undernourished, and 33 percent of African children are stunted, underweight, or wasted (FAO, 2004). Hunger, poverty, disease and vulnerability to environmental change are all closely linked. Africa has remained the only major developing region showing a decline in income per capita over the period 1980–2000 (Sachs *et al.*, 2004). Two-thirds of the least developed countries are in Africa. Additionally, Africa's *health conditions* are the worst on the planet (Sachs *et al.*, 2004). It has the highest under-5 mortality rate (140 per 1000) and life expectancy at birth is only about 54 years.

Armed conflicts: Threatened lives and livelihoods increase the vulnerability to natural hazards and disease, and spawn refugees who degrade the environment for lack of other resources.

Lack of education: Over 60% of Africa's adult population is illiterate (FAO, 2004).

Desertification and Deforestation: Dry lands cover 43% of the continent and a third of the population lives in them (World Water Forum, 2000). West Africa is the source of nearly half of the mineral aerosols entering the atmosphere globally (Andreae, 1995; Duce, 1995).

In Africa, this area amounted to 5.2 million ha/year (0.78 % of the forested area per year) for the period 1990–2000. The multiple causality of environmental change offers the opportunity to examine multiple approaches and frameworks of knowledge construction (Airhihenbuwa, 1995; Carruthers, 2005; Oomen, 2005) and methodologies of analysis (Freire and Faundez, 1989), particularly since Africa is so culturally diverse. The climate of Africa is diverse and controlled by complex interactions between the oceans, land and atmosphere at local, regional and global scales. On average, Africa is hotter, drier and has less-dependable rainfall than most other regions of the world. As a consequence, and considering the high climate dependency at all levels from the individual household to the regional economy, several studies have concluded that Africa is among the most vulnerable continents to climate change and variability (Fischer *et al.*, 2002 and 2005; IPCC 2001).

Industrialization: Africa is causing a lot of emission of carbon dioxide. At the current rate of emission of carbon dioxide and other greenhouse gases, the average temperature of the planet could increase by as much as 4°C from its preindustrial temperatures by the end of this century. The concentration of carbon dioxide (CO) in the atmosphere is about 385 ppm (parts per million) and increasing at an alarming rate of 2 ppm per year.

Water Supply: Except in the equatorial region and coastal areas of eastern and southern Africa, the continent is dry, sub-humid to arid. Global warming will result in a reduction in soil moisture in sub-humid regions and a reduction in runoff, because high temperatures enhance evaporation. Indeed, World Water Forum (2000) reported that the southern Africa region will experience the greatest increase in runoff by the year 2050, increasing the number of countries included in the water stress category (using a per capita water-scarcity limit of 1,000 cubic meters per year). As an additional threat to African development, access to water is likely to be the source of an increasing number of conflicts in the future; as have been observed already between Senegal and Mauritanian during the late 1980s and early 1990s (World Water Forum, 2000).

The Key elements and the Way Forward

Land Degradation

Land degradation is defined as a persistent decrease in the capacity of the land to deliver ecosystem services. Degradation of dry lands is called 'desertification', which has been a concern in Africa for decades. The lack of progress in solving the issue of land degradation in the region is partly because the underpinning science has not received sufficient attention and there is little consensus on the causes, consequences or best remedies for land degradation in general and desertification in particular. There are both local and global causes of land degradation – for instance, circumstantial evidence linked the sharp drop in Sahelian rainfall in the 1970s to changes in the global ocean and atmosphere circulation. There are regional and global consequences to land degradation, including impacts on the climate system through changes in emissions of greenhouse gases and aerosols and surface albedo, loss of biodiversity, and the displacement of people who no longer are able to sustain livelihoods in the degraded lands.

Biodiversity loss and human well-being

Although Africa does not have a disproportionate share of the world's original biodiversity, it has retained a larger fraction than many other regions (Scholes *et al.*, 2006). For this reason, non-Africans travel to Africa for nature-based tourism and lobby for the protection of African forests, wildlife and coastlines. Africans have always had a direct and intimate dependence on natural resources and biodiversity, but the mechanisms which have allowed their past coexistence have broken down under the pressures of the modern era. Loss of biodiversity, at all levels from the gene to the ecosystem, has a direct and negative impact on the well-being of Africans (Scholes *et al.*, 2006). Many natural resources, including fish, firewood, timber, medicinal plants and wild foods are increasingly becoming scarce. The causes of accelerating biodiversity loss vary between locations and between the major plant and animal groups: over-harvesting has been an important historical cause of fisheries, forest and wildlife decline; but at present habitat loss, fragmentation and degradation is the key threat.

Climate change is projected to be the dominant driver of biodiversity loss by the middle of the 21st century (Scholes *et al.*, 2006). Biodiversity is important to Africans for three reasons: the biodiversity of Africa at all scales is high and relatively intact but threatened; African peoples have a close relationship with and high dependence on biodiversity; and biodiversity is a relative strength of African researchers. Biodiversity research in Africa must go far beyond the study of wild species in protected areas. It should be people-oriented, with a focus on the sustainable use of biodiversity-based resources and the persistence of viable levels of biodiversity at the gene, species and ecosystem levels within terrestrial, freshwater or coastal ecosystems that are inhabited and used by people. Biodiversity loss and its effects on human well-being is a key issue on land, in freshwaters, on the coast and in the oceans. On the land, it is not only concentrated in tropical rainforests or biodiversity hotspots, but also in the dry lands. Freshwater biodiversity in Africa, as on other continents, is often the first to be transformed by human activities, through over-fishing, water abstraction, sedimentation, pollution and the invasion of alien species. Marine biodiversity, which underpins economically and nutritionally-important fisheries, is a critical concern around the African coast. It is threatened by the direct and indirect effects of fishing, coastal pollution and climate change (particularly with respect to coral bleaching). The cross-boundary nature of all ecosystems is important, but especially so in the fluid context of marine ecosystems.

Rainfall in Africa

The most crucial climate issue for decision-makers at all levels, from the household level to the African Union, is rainfall. Unfortunately, it is also the variable associated with the greatest uncertainty in climate projections. For many regions in Africa, there remains disagreement about even the direction of future rainfall trends, as well as the magnitude. Even at the scale of seasonal rainfall forecasts, which are critical for food and water security in the semi-arid regions of Africa, wide uncertainty persists. Since improved adaptation to inter-seasonal variability in rainfall is thought to be a useful analogue for enhanced capacity to respond to decadal climate change, the need to increase the confidence in the medium-term forecasts is an important and practical objective. Most rainfall in the affected areas of Africa is in the form of high-intensity convective storms. Prediction of such events on a case by case basis is inherently difficult; because of their complex dynamics and relatively small scale, very little has been done on this aspect because of the global distribution of meteorological researchers. The key objective of such research is not to achieve prediction for individual storms, but to provide an accurate statistical picture of storm probabilities, rainfall per storm and extent.

Land-use and cover change in Africa

In Africa, interactions with the land are particularly important because of their impacts on biodiversity, including that of disease-bearing organisms; on biogeochemical interactions, air quality, water yield and climate; and on the capacity of the land to sustain livelihoods. There is tantalizing evidence that convective rainfall patterns at a regional scale may be partly linked to land surface conditions, through mechanisms such as the changes in surface albedo that result from reduction in vegetation cover, and the influence of dust and smoke aerosols on cloud condensation nuclei. If these feedback mechanisms are significant, they have important consequences for land use policy.

Resilience of food supply systems

Reducing the vulnerability of Africa to global climate change requires building the resilience of Africa's socio-ecological systems, particularly through achieving food security in the region. Food security is achieved when all people, at all times, have physical and economic access to sufficient, safe and nutritious food to meet their dietary needs and food preferences for an active healthy life (FAO, 2004). Policy interventions to improve food security in the region need to take global environmental change into account. Africa's food insecurity arises from a combination of factors such as poor agricultural infrastructure, a rapidly growing young population, widespread poverty, war and conflicts, poor governance, a heavy disease burden and unfavorable soils and climate conditions among others. Integrated research needs to be carried out at three levels: first, at the household and community levels, where vulnerability is usually more clearly expressed, second, at the sub-regional level, so that the interaction of policy and food systems' vulnerability can be addressed and finally, at the continental level, to understand how findings can be scaled up and outwards for policy and wider use. Here engagements with practitioners and those actively working on food systems, from various perspectives, may enhance this agenda.

Water resources and their governance

Africa is the second driest continent, after Australia, but unlike Australia, it supports over 800 million people, many of whom live in dry lands. Water-imposed constraints on human development, including absolute shortage, declining water quality and water-borne diseases are already a reality in Africa, and will intensify in the future. The volume of stream-flow and its split between reliable base-flow and damaging storm-flow are influenced by changes in the climate and on land surfaces, as well as by impoundments and abstractions of water for human use. A large fraction of African people depend on groundwater, which is used unsustainably, virtually everywhere. The decline of water resource quality is due to land-and sea-based human activities such as the disposal of domestic sewage and industrial waste water, toxic leachates from mining, and the effects of agricultural runoff containing sediments, nutrients and pesticides. Water quality and quantity interact through the loss of dilution and flushing capacity. The impacts accumulate down the drainage basin and have serious consequences for the coastal zone, where they are compounded by maritime activities. These changes have consequences for fresh water availability for domestic, agricultural and industrial use, food and employment security, and loss of human life and property. Solutions to these multi-faceted problems require a multi-disciplinary approach. In particular it needs the greater involvement of social scientists, with experience in policy analysis of problems where environmental, social, political and economic variables interact. Progress towards more sustainable water resource management will require the establishment or strengthening of research and development programs at the national, sub-regional, regional and international levels. Key research issues include:

(a) Integrated water resources development and management (including Protection of water resources, water quality and aquatic ecosystems; Drinking water supply and sanitation; Water and sustainable urban development; Water for sustainable food production and rural development; and Impacts of climate change on water resources). (b) Research and development of appropriate technical solutions. (Increase urban treatment capacity commensurate with loads; Identification of critical areas for water-related adaptive research. Strengthening the adaptive research capacities of institutions; Enhancement of translation of research results on water-related farming and fishing systems into practical and accessible technologies) and (c) Strengthening of sector monitoring and information management; Pursuit of low-cost scientific and technological means, as far as practicable; Utilization of traditional and indigenous practices, as far as practicable; Assistance to country-level technical/scientific institutes to facilitate curricula development to support fields critical to the water and sanitation sector and treatment and safe re-use of domestic and industrial waste waters in urban and rural areas).

There is an emerging consensus that wise water management cannot be accomplished by technical or regulatory measures alone, but must encompass education and awareness-raising initiatives.

Environmental and atmospheric pollution

Africa, like other developing continents, is undergoing rapid urbanization and industrialization with the consequent problems of land, water and air pollution, particularly in urban areas. These problems are compounded by poverty, lax regulatory policies and obsolete mitigation technologies and may be exacerbated by rising global temperatures and increasing atmospheric burdens of dust and smoke resulting from burning of biomass. Observations and models show that the tropospheric concentrations of most reactive chemical species are increasing in time and space over all of tropical Africa. In developing countries, sulphur dioxide, nitrogen compounds, ozone and particulate matter are the main hazards. Warm temperatures with high relative humidity and a high frequency of precipitation increase the potentials of corrosion damage, especially that caused by wet deposition. Global – and particularly African – emissions of ammonia are rapidly increasing and are believed to constitute one of the major future environmental problems due to their acidifying and eutrophication potentials.

Capacity building

Capacity building is a challenge in virtually all countries, but there are peculiar issues in developing countries, where limited financial resources and near total absence of universal education systems amplify the problems. There is wide and growing acceptance internationally of the need to build capacity for adaptation to global climate change, particularly in the more vulnerable parts of the world. Although climate change is a global environmental issue, neither the industrial causes of the problem nor the tools of its analysis and mitigation are native to Sub-Saharan Africa. As a result, Africa is faced with the necessity of dealing with a complex economic, scientific and social issue, having neither the scientific background, the necessary tradition of public awareness and mobilization nor having the availability of financial resources and implementation structures to do so. Capacity building here means providing frameworks for project identification, formulation and implementation and making the maximum use of existing skills and resources.

The following limitations will need to be addressed if scientific and technical capacity is to be built and sustained: (a) Scarcity of scientists, science infrastructure and dwindling funding for research and development; poor capacity of existing institutions, limiting local, regional and national authorities and civil societies to respond to disasters, and to adapt to global change and to plan and adopt mitigation measures; poorly coordinated involvement of all sectors/stakeholders of society in recording information on climate and natural resources; poor capacity to assess vulnerabilities and adaptation strategies, especially using community-based techniques for measurement and recording of information and assessing vulnerability to global environment change, and the training of researchers to better assess the risks faced due to these changes.

Others include: Integration of adaptation responses into national development strategies building capacity to respond to climate change events. This includes reducing the risk of climate impacts on infrastructure investments, enhanced disaster preparedness, use of traditional knowledge and new technologies and coordinated regional response; Enabling environments for technology transfer – introduction of appropriate technologies to aid adaptation (flood control systems, emergency shelters, etc.) and mitigation (energy efficiency, reforestation, etc) and the strengthening capacity to adopt and maintain new technologies; Lack of high level political commitment, which leads to financial instability as predictable resourcing is not provided for in central budgetary processes; poor Communication difficulties between the agencies, institutions, government departments, NGOs or community groups involved in the capacity building activity; Complex institutional policies including overly-bureaucratic systems and difficulties in identifying key advisors for training opportunities; and Difficulties in targeting the most appropriate candidate for training activeties and difficulties in reaching people outside capital cities. Capacity building includes not only training and developing the next generation of scientists (i.e. human capacity), but also ensuring that the institutional capacity is enhanced in Africa. This requires a multi-faceted, coordinated and deliberate effort, including creating institutions for advanced study and training supportive networks in areas such as modeling; engagement of non-African experts in local training; improving the retention of skills in Africa; creating incentives for temporary or permanent participation of African experts currently abroad; encouraging regional mobility of experts and mutually beneficial north-south and south-south cooperation, and developing receptive capacity in the end-users of research and the research communicators (extension workers) who link together all the component units.

Capacity building activities in the public sector should be an integral part of overall public sector reform processes. In the interest of sustainability, it is important that responses to risks to the environment be mainstreamed within nationally-owned strategies such as poverty reduction strategies and into existing projects and programs relating to the range of government ministries. Specific capacity building projects are more successful when they establish policy links to other ministries such as agriculture, water, energy and finance. These, combined with the external financial and technical support are usually required to achieve sustainable results. Institutional capacity building should involve decision-makers at the highest level. This is necessary to ensure continuous support for capacity development after the initial program is completed. In local and regional programs it is essential to have appropriate regional and national authorities and institutions on board to assure full acceptance of and support for the initiative. Both donors and host countries must adopt a long term approach to capacity development and this requires financial sustainability, ultimately supported by national policies and budgets that reflect national policy priorities.

The science-practice-policy interfaces

The pressing development needs in Africa make 'science for science's sake' an inappropriate and wasted opportunity. While there is clearly a need for undertaking fundamental and science-led research, the realities of the continent require a very pragmatic approach to addressing the real-world problems and setting our evidence-based priorities. A well-functioning set of science-practice-policy interface is needed therefore to ensure that what is known is swiftly negotiated, debated and efficiently applied (Guston, 2001). Such interactions between science and society are better described as spider webs rather than unidirectional pipelines or even bidirectional bridges. The diversity of power relations between actors is a key issue (Miller, 2001); a range of 'brokers' and institutional designs that can both convey and receive information between producer and user groups are suggested (Jasanoff, 1987; Cash, 2001; Guston, 2001; Miller, 2001; Cash and Moser, 2000; Vogel *et al.*, 2006). The products of scientific research need to be credible, salient and legitimate (Cash, 2001). This means that in addition to greatly improving the dissemination of results after the research is done, the research agenda, where relevant and appropriate, need to be negotiated with users at the outset.

References

Airhihenbuwa, C. O., 1995. Health and culture beyond the Western paradigm. Sage Publications, International Educational and Professional Publisher, Thousand Oaks, London and New Delhi.

Andreae, M. O., 1995. Climatic effects of changing atmospheric aerosol levels. World Survey of Climatology 16, Future climates of the world. A. Henderson Sellers (ed.). Elsevier Amsterdam, 341–392.

Benson, T., 2004. Africa's food and nutritional security situation: Where we are and how did we get here. International Food Policy Research Institute, 2020. Discussion Paper, 37.

Carruthers, J., 2005. Africa's environmental history. IHDP Update 02/2005 www.ihdp.org/updatehistory05/references.htm.

Cash, D. and S. Moser, 2000. Linking global and local scales: Designing dynamic assessment and management processes. Global Environmental Change, 109- 120.

Cash, D., 2001. In order to aid in diffusing useful and practical information: agricultural extensions and boundary organizations, Science Technology and Human Values Special Issue: Boundary Organizations in Environmental Policy and Science, 431–453.

Duce. R., 1995. Sources, distributions and fluxes of mineral aerosols and their relationship to climate. Pro Workshop on Aerosol Forcing of Climate Charlson and Heintzenberg (eds.), Berlin, 43–72.

FAO, 2004.The implementation of global plan of action for the conservation of plant resources for food and agriculture commission on genetic resources for food and agriculture. Tenth Regular Session Rome, 80–12, Nov. 2004.

Fischer, G., Shah, M. and H. van Velthuizen, 2002. Climate Change and Agriculture Vulnerability, International Institute for Applied Systems Analysis, Austria.

Fischer, G., Shah, M., Tubiello, N. and H. van Velthuizen, 2005. Socio- economic and climate change impacts on agriculture: an integrated assessment, 1990-2080. Phil. Trans. R. Soc., 360 2067–2083.

Freire, P. and A. Faundez, 1989. Learning to Question. New York Continuum. Gibbons, M., Limoges. C., Nowotny, H., Schwartzman, S., Scott, P. and M. Trow, 1994. New Production of Knowledge, the Dynamics of Science and Research in Contemporary Societies, 7/7e. SAGE Publications, London.

Guston, D., 2001. Boundary organizations in environmental policy and science: an introduction. Science Technology and Human Values Special Issue: Boundary Organizations in Environmental Policy and Science, 26, 4 399–408.

IPCC, 2001. Climate change 2001 – Impacts adaptation and vulnerability, Cambridge University Press, Cambridge.

Jasanoff, S. S., 1987. Contested boundaries in policy relevant science. Social Studies of Science, 17 195–230.

MEA, Millennium Ecosystem Assessment, 2005. Ecosystems and human wellbeing: Current state and trends, Volume I–IV, Island Press, Washington.

Miller, C. 2001. Hybrid management: Boundary organizations. Science policy and environmental governance in the climate regimes. Science Technology and Human Values Special Issue: Boundary Organizations in Environmental Policy and Science, 26 478–500.

Omen, B., 2005. Chiefs in South Africa law power and culture in the post- Apartheid era. James Currey, Oxford.

Sachs J. D., Mcarthur, J. W., Schmidt-Traub, G., Kruk, M., Bahadur, C., Faye, M. and G. McCord, 2004. Ending Africa's Poverty Trap, Brookings Papers on Economic Activity, 1 117–240.

Scholes, R. J., Kuper, W. and R. Biggs, 2006. Biodiversity, Africa Environment Outlook 2 – Our Environment, Our Wealth (AEO-2), Chapter 7. UNEP in press.

Vogel, Moser, Kasperson and Daebelko, 2006. In Review GEC. World Water Forum, 2000. The Africa Water Vision for 2025. Equitable and sustainable use of water for socio-economic development. WWF, The Hague, 30 pp.

Cyprian Osinde
Makerere University, Uganda
Joseph Okello Onen
Gulu University, Uganda

Adaptation measures taken to preserve animal species affected by climate change in Karamoja sub-region, Uganda

Abstract
The possible impacts of climate change on food security and human health are of increasing concern especially in African regions with pre-existing systemic risks such as arid climate, poverty and environmental degradation. Extensive areas of Sub-Saharan Africa and especially the Eastern flanks are considered to be particularly vulnerable to both short term increase in climate variability and concomitant species extinction and long-term consequences of climate change. Increasing the resilience of existing livelihoods to such climate changes is considered as a key strategy for adaptation. The project developed and enhanced the resilience of poor rural communities to climate change by specifically addressing linkages between climate variabilities, food security and human health in agriculture-based livelihood systems. Based on a detailed assessment of trends in climate variabilities and its direct and indirect impacts on livelihood security, the project identified and enhanced community-based capacities and policy frameworks favorable to preventing species extinction. In this connection, the project initiated multi-stakeholder action-research processes in different community settings, each of which designed and experimentally tested pilot interventions. The insights from applied research strengthened multi-stakeholder development processes, supported adaptive livelihoods systems, and informed natural resource management related policies which led to species preservation decisions in Karamoja region.

Introduction

There is a growing body of evidence that global climate is changing (Parry *et al.*, 2007). The changing trend of climate poses major challenges to the region of Africa south of the Sahara desert – the Sub-Saharan Africa. These challenges must be addressed through comprehensive Africa-centered initiatives involving people, communities and relevant research. Climate change refers to a change in the state of the climate that can be identified (e.g. using statistical tests) by changes in the mean and/or the variability of its properties, and that persists for an extended period, typically decades or longer (IPCC, 2007b). Climate variability on the other hand refers to variations in the mean state and other statistics (such as standard deviations, the occurrence of extremes, etc.) of the climate on all spatial and temporal scales beyond that of individual weather events (IPCC, 2007b). A recent report of the Intergovernmental Panel on Climate Change (IPCC) concluded that "*warming of the climate system is unequivocal, as is now evident from observations of increases in global average air and ocean temperatures, widespread melting of snow and ice, and rising global average sea level*" (IPCC, 2007a; Parry et. al., 2007). The variability in climate and the frequency of extreme weather events have increased significantly in recent decades. This has and will continue to cause major disruptions in the functioning of ecosystem health and services leading to negative impacts on the well being of their constituents (MEA, 2005). The negative impacts will be felt most severely by people in poor countries of Sub-Saharan Africa (Chancellor *et al.*, 2008).

Uganda will experience a pronounced impact of global climate variability and will be among the countries in Sub-Saharan Africa hardest hit by the immediate impacts of climate variability including species extinction. It is estimated that Uganda will be 80% desert in 100 years time if current trends continue and this will result in species extinction (IPCC, 2007a). The country has human population estimated at 30 million people. Agriculture is the mainstay of the economy – providing a significant share of GDP (about 40%), export earnings (85%) and total employment (80%). In addition, animal husbandry is a considerable source of employment and livelihood in the country, representing 7.5% of the GDP and 17% of the agricultural GDP (Uganda Bureau of Statistics – UBOS, 2005). Most (85%) of the livestock production in the country takes place in the rangelands of cattle corridor, which runs from the south-west to the north-east direction, from the Rwanda border to the Sudan and Kenya borders, and covers an estimated area of 84,000 km^2 constituting 44% of the country's total surface land area. In addition, the corridor produces a significant amount of biomass used as energy sources (firewood and charcoal) by communities including those outside the area. According to Ziervogel *et al.* (2005), rangeland condition is directly affected by the climate and, in turn, directly affects the quality and quantity of small and large stock and associated livelihood activities (Fig. 1).

Fig. 1: Effects of climate variability on cattle in Karamoja region.

Uganda's climate is naturally variable and thus the country is very susceptible to flood and drought events which have had negative socio-economic impacts hurting the economy and well-being of its people (DFID, 2008). This variability had caused and will continue to cause major disruptions in the functioning of ecosystem health and services (MEA, 2005), thus exacerbating poverty and triggering migration as well as heightening the already existing stiff competition over strategic water resources. Climate change could lead to regional social insecurity (DFID, 2008). The Karamoja region suffers from severe climate variability and this has shaped the way of living of the communities (NAPA, 2007). The region is severely degraded by a number of factors ranging from overgrazing, severe drought, outbreak of pests and diseases, bush burning, fuel wood harvesting, insecurity and poor research and extension programs all leading to animal species extinction (NEMA, 2006; Keizire and Mugyenyi, 2006; National land policy, 2007; NAPA, 2007). Currently, Karamoja region is experiencing unusually long dry spells, with grass withering and water sources either drying or water volumes receding every year, soils becoming increasingly unproductive and resulting in negative impacts on animal and agricultural production (NAPA, 2007). These have led to famine, malnutrition, serious health conditions and frequent migration, and loss of biodiversity. In order to look for pastures, water and food, most of the communities have adopted a nomadic lifestyle over the last 2 decades which makes them roam throughout the region and over large areas in districts neighbouring the region. The practice has brought about cattle rustling, conflict with neighbours and insecurity of all sorts.

Karamoja region is in a state of chronic hunger, malnutrition and diseases as a result of drought and water scarcity. Livestock die in large numbers annually. In addition, there have been failed harvests, and outbreaks of crop and animal disease epidemics, leaving several million people food insecure and disease prone. According to Olive Olupot, who had been living in this region for long, "climate change is a matter of life and death". This is accentuated in the following statement: "*We suffer a lot of pain, because we can no longer feed our children. To make matters worse, many local people are turning to cutting down trees for charcoal in order to earn money as a way to survive. This is worsening the problem. With fewer trees, the rainfall patterns are likely to become more unpredictable*" (New Vision, 2009).

The existing national and community-based adaptation initiatives and policies to climate change have not been fully implemented to mitigate the negative effects of climate variability. In addition, community-based adaptation initiatives and supportive policies for climate change mitigation have neither been characterized nor properly understood. Worse still, the local capacities of individuals, communities and institutions to tackle the problems associated with climate variability are not well evaluated nor are they fully exploited. The past and current efforts by the Government have tended to address these problems piece-meal and in a superficial manner. The goal of the study was to develop and enhance community-based climate variability adaptation initiatives and supportive policy measures, which specifically improve food and health security in the region on a long-term perspective, hence coming up with strategies which preserve species from extinction. It explored the past and current community-based livelihoods' resilience and adaptation capacities, strategies and options to the effects of climate variability on food and health security in the region and hence coming up with protective measures to prevent species extinction due to climate change.

This study is in response to the call by the global community (The Kyoto Protocol, 1997; WHO, 2008) for urgent actions and stronger commitment of member states to protect life and health from negative impacts of climate change, including species extinction. The study built up capacity for different categories of stakeholders, particularly the boundary partners (individuals, communities and institutions in the study area where change/appropriate adaptations are desired), as it was tailored to sustainable creation, utilization and retention of capacity to effectively manage risks arising from climate change, enhance self-reliance and improve the lives of communities.

Methodology

This study was formulated in collaboration with the community leaders in Karamoja region: consultative meetings, stakeholder analysis and community diagnosis exercises were undertaken to identify the essential stakeholders and characterize the communities in the study. Different gender categories were involved in the implementation of the research activities and collection of data using household surveys. The communities contributed in adaptation of strategies for climate change mitigation, including adaptation strategies for prevention of animal species extinction.

Results and discussion

The knowledge and information about livelihood resilience and adaptive capacity to prevent extinction of species due to climate change was availed by the community members themselves. For example, the community members reported to have learned protective strategies to prevent species extinction due to climate change. Communities have learned to put up a net work of nature wild life reserves, where important plant species which were affected by climate change could be preserved by planting them. In case of animals, the communities have adapted a nomadic form of lifestyle should harsh environment conditions arise? The people tend to move from place to place in search of water and pasture to prevent starvation of animals in the region.

To enhance adaptation capacities and resilience to climate change, this community knowledge needs to be harnessed, sharpened and utilized for the betterment of the communities. This is best done along the continuum of applied knowledge acquisition (LDG, 2004), starting with (i) Derivation (information derived from local context to meet a local need) through (ii) Relevance (the communities seeing the information relevant and usable); (iii) Delivery (knowledge accessed by the communities); (iv) Uptake (correct application) and ending with (v) Action (changing behavior = adaptations). Other information obtained from the communities included: Village resources maps which were used to identify the environmental patterns such as water sources, forest cover, farming land, economic and social resources in the community. These resources were used in a controllable manner during times of excessive drought, to prevent death of livestock. Farming was one of the protective strategies which were being used by the communities to prevent extinction of wild species of plants in addition. Some plants could easily perish in the wild, and for that reason the susceptible plants were brought into cultivation by the communities, hence helping in preventing their extinction. Other resilient forms of livelihoods which enabled the individuals and communities to prevent extinction of species due to climate change included: anticipation, surveying, forecasting, developing preventive measures to increase tolerance, recovery and restoration, and learning processes (Conway and Waage, 2010).

Policy analysis

Policy analysis was undertaken, involving the essential stakeholders and community leaders, to identify and review the current policies that purportedly support communities to adapt to preventive measures which prevent species extinction owing to climate change. In Uganda, a number of policies have been put in place. However, the communities and other relevant stakeholders at local level may not understand these policies, and the local capacity to implement them may be limited. Hence, there is need for the implementers of these policies to come up with implementing strategies which will prevent communities to over exploit the environment leading to preservation of species. Arising from this, a capacity building plan was developed to enhance community livelihood resilience and adaptation capacities, strategies and options. The plan was tailored to sustainable creation, utilization and retention of capacity to effectively manage risks arising from climate variability, enhance self-reliance, and improve people's lives and at the same time preventing extinction of species due to climate change.

Conclusion

For the primary beneficiaries, the impact of the project is cost-savings on nomadic lifestyle, adoption of appropriate adaptation strategies and reduced competition for resources. Hence, enhanced income of farming communities through increased productivity and sales of animals and their products will be the consequence. In addition, the communities benefited from introduced skills for adapting to the impacts of climate change, as compared to other farming communities outside the study areas. Overall, the project had positive effects on the environment by reducing intensity of competition for resources and land degradation. Increased utilization of technologies for adapting strategies mitigation of climate change also improved biodiversity of species that were threatened with extinction.

References

Chancellor, T., Morton, John and Tabo Ramadjita, 2008. *Climate change and agricultural development.* Draft SCARDA Briefing Paper. Natural Resource Institute, University of Greenwich, Chatham. 12 pp.

Conway, G. and J. Waage, 2010. Science and innovation for development. UK Collaborative on Development Sciences (UKCDS), 380.

IPCC, 2007a. Climate change 2007: Impacts, adaptation and vulnerability. Working group I. Contribution to the Fourth Assessment Report of the Intergovernmental Panel on Climate Change. Summary for Policy Makers and Technical Summary. ISBN92-9169-121-6.

IPCC, 2007b. Climate Change 2007: The physical science basis. Working group I. Contribution to the Fourth Assessment Report of the Inter-governmental Panel on Climate Change. Summary for Policy Makers, Technical Summary and Frequently Asked Questions. ISBN92-9169-121-6.

DFID Uganda, 2008. Climate Change Scoping Mission.

Keizire, B. B. and O. Mugyenyi, 2006. Mainstreaming environment and natural resource issues in selected government sectors. ACODE Policy Research Series, No. 21.

Millennium Ecosystem Assessment, 2005. Report of the Millennium Ecosystem Assessment 2005. Ministry of Lands, Housing and Urban Development, Uganda, 2007. National Land Policy, 2007.

National Adaptation Plan of Action, 2007. Report on climate change: Uganda National Adaptation Plan of Action, 2007.

NEMA, 2006. Report on ecosystems, ecosystem services and their linkages to poverty reduction in Uganda.

New Vision, Vol. 24, no. 110, Friday, June 5, 2009.

Parry, M., Canziani, J., Palutikof, P., van der Linden, J. P. and Clair Hanson, 2007. Climate change: Impacts, adaptation and vulnerability. Contribution of Working Group II to the Fourth Assessment Report of the Intergovernmental Panel on Climate Change. Cambridge University Press, 79 pp.

The Royal Society, "Reaping the benefits. Science and the sustainable intensification of global agriculture. RS Policy Document 11/09." Royal Society, London, 2009.

The Kyoto Protocol, 1997.

Uganda Bureau of Statistics, 2005. Report of Uganda National Household Survey, 2005/6.

WHO, 2008. WHO report protecting health from climate change: Global Research Priorities (in press). Check at http://www.who.int/globalchange/climate/EBCC health resolution/en/index.html for link to report.

Ziervogel, G., Nyong, A., Osman, B., Conde, C., Downing, T. E. and S. Cortés, 2005. "Climate variability and change: Implications for household food security." In: Leary, N., Conde, C., Nyong, A. and J. Pulhin (eds.). For whom the bell tolls, case studies of climate change vulnerability. AIACC Working Papers. (http://www.aiaccproject.org).

Etiosa Uyigue
University of Benin, Nigeria

Physical and economic consequences of climate change in select-countries of Sub-Sahara Africa

Abstract

Globally, climate change is increasingly emerging as a serious threat to economic growth, social well-being, quality of life and political stability. The African continent is particularly vulnerable to the various manifestations of climate change for many reasons: widespread poverty; reliance on biomass and traditional energy sources; recurrent droughts; inequitable land distribution; and over-dependence on rain-fed agriculture. Climate change phenomenon comes with it increases in temperatures, decline in rainfall amounts, increases in sea levels, uncontrollable coastal erosion, unpredictable seasons of years, desert encroachment and flooding. Creating awareness among local people, concentrating development efforts in rural communities in Africa and promoting alternative sources of energy and energy sovereignty are recommended for the mitigation of the problems posed by climate change at the local level. At the regional level, the collection, synthesis and analysis of climate data for the purpose of forecasting climatic changes is desirable. Building the capacity of African scientists and policy makers and integrating many disciplines, methods of analysis and strategies in tackling the menace are recommended also.

Introduction

Climate change can be defined as the variation in climate over a period of time ranging from few years to decades. It can also be defined as the permanent departure of climate patterns from mean values of observed climate indices. When rainfall amount, for example, in any particular month, season or year deviates from what has been normally observed, the departures are noted. The observed variation, if noted from year to year, is an example of climate variability. When the departure persists, it is technically referred to as climate fluctuation. If the fluctuations persist for a sufficiently long time and is further found to be statistically significant, then this is an indication that climate change has occurred. Human activities both in agricultural and industrial sectors are causing the emission of greenhouse gases into the atmosphere leading to the general warming of the lower atmosphere. The warming atmosphere interacts with other climatic components and alters climate at any point. Some of the effects and impacts of climate change include: severe droughts, accelerating desertification, drying up of wetlands, shrinking lake basins, and changing ecosystems; extreme fluctuations of river flow, reduced crop yields; expanding zones of tropical vector diseases; and rising sea level and coastal flooding. Globally, climate change is increasingly emerging as a serious threat to economic growth, social well-being, quality of life and political stability. It is clear that the world's poorest countries are expected to be the least likely to adapt to climate risk due to geographic location, low incomes and weak institutional capacity. It is one of the most alarming challenges facing African development efforts.

In the climate change scenario, the African continent is particularly vulnerable to the various manifestations of climate change for many reasons. A large percentage of the African population is poor – over 40% of the people in Africa live on one Dollar a day or less (O'Brien and Leichenko, 2005). Moreover, deterioration in terms of trade, inappropriate and insufficient policies, high population growth rates, and lack of significant investment will further exacerbate the impacts of climate change and undermine development efforts in Africa. At the same time, Africa continues to suffer from a significant deficit in infrastructure, with about 60% of African countries affected by the energy crisis and overall access to electricity at only about 24%. Moreover, Sub-Sahara Africa continues to rely heavily on biomass and traditional energy sources. This over reliance on traditional energy sources leads to low level of energy efficiency, deforestation and biodiversity loss, greater health hazards due to indoor air pollution and reduced capacity to mitigate climate change.

The reports of Hulme (1996) and the IPCC (1998) concluded that the African continent is particularly vulnerable to the impacts of climate change because of factors such as widespread poverty, recurrent droughts, inequitable land distribution, and overdependence on rain-fed agriculture. While the natural resource sector is the development engine in Africa, it is also the most climate-sensitive sector. Agriculture in Africa remains rain-fed; only 7% of agricultural land is currently under irrigation (Harmeiing et al., 2007). Yet agriculture contributes approximately 30% to GDP and employs about 70% of the African population. Many of the poor live in rural areas and their lives depend on natural sectors such as farming and fishery. It therefore means that rain-fed agriculture is essential for agricultural development in the region and changes in rainfall patterns are therefore extremely significant.

Further, two thirds of the region is desert or dry land and Africa already has the highest mortality related vulnerability coefficients for drought. In Africa, dry lands cover 1959 million hectares of land in 25 countries making up 65% of the continent and one-third of the world's dry land. The pressure on land in many of these countries is already great and climate change will worsen the trend and bring greater hardship to people through the effects of severe drought (Imevbore, 1995).

The objectives of this research are to highlight the various observations on climate change that have been made in different parts of Sub-Sahara Africa, the current and future problems associated with climate change and proffer possible solutions on how these problems could be tackled at local and regional levels.

Climate change observations and problems in Sub-Saharan Africa

Temperature

Observational records show that the continent of Africa is warmer than it was 100 years ago (IPCC, 1996). Warming through the 20th century has been at the rate of about 0.05°C per decade with slightly higher warming in the June, July, August and September-November seasons (Hulme *et al.*, 2001). The 5 warmest years in Africa have all occurred since 1988, with 1988 and 1995 being the two warmest years. This rate of warming is not dissimilar to that experienced globally; increases in temperature will create an imbalance in climatic systems and this will lead to extreme climate conditions.

Change in rainfall pattern

In Africa, one of the most significant climatic variations has been the persistent decline in rainfall. In the Sahel region, this decline has been experienced since the late 1960s (Lamb and Peppier, 1992). Mean rainfall decreased by 20–49% in the Sahel between the periods 1931–1960 and 1968–1997 and generally 5–10% across the rest of the continent. The impacts of the Sahelian drought aggravated the already stressed food security situation in many countries in Africa prompting the various levels of government to embark on a rigorous policy to increase food production. To achieve this, impoundment of river basins was seen as inevitable to provide sufficient water for year-round irrigation, which led to the construction of over 246 dams in Nigeria (Imevbore *et al.*, 1986). More recently, the pattern has been one of increased aridity throughout most of the continent. The persistent drying trend has caused concern among development planners regarding how to cope with losses of food production, episodes of food insecurity, displacements of populations, lack of water resources, and constraints on hydroelectricity.

Many communities in Africa practice rain-fed agriculture. Decreasing precipitation will adversely affect food security in the African continent. It will also lead to a decline in water availability for people and livestock and for sustaining the wider environment. Because of the change in rainfall pattern, farmers who plant after the first or second rain run into huge losses when the rains are delayed beyond the usual due to climatic change. The crops are scotched causing huge economic losses.

Reports from Burkina Faso revealed that the amount of rain has reduced and the rainy season has become very hot. In Malawi, rainfall has become very erratic and there is less each year. The streams and rivers are drying up and the water table is now deeper than before. Reports from Mali revealed that climate change has affected water security. The rains have become very scarce and there is very little rise in water level. Because of this change, many have to walk several kilometers to look for water. The changes have caused forests to dry up and games have disappeared (Tearfund, 2005). In Mozambique, many wells that have always contained water are now dried up making many communities to live in great difficulty. Many have to walk distances of 10–30 kilometers to fetch drinking water. Some do not wash every day in order to save water. In the past in Niger, the rains came for three months at a time and the temperature was quite cold. Now it only rains for two months or less; this trend started after 1984. There has been severe drought leading to the death of many animals (Tearfund, 2005).

Ethiopia has experienced climate change that adversely affected water availability. The streams are drying up and there is decreased flow in rivers, and lakes are shrinking affecting the habitat of birds and fish. Still in Ethiopia, ground water levels are getting lower, there is reduction in the volume of water for hydroelectricity plants. In the Rift Valley of Ethiopia the use of water for soda ash extraction from Lake Abiyatta has caused the lake water to recede by as much as 3–4 km compared to what it was 30 years ago (Tearfund, 2005). In Nigeria, the decline in rainfall in Nigeria started at the beginning of the 1960s when a decade of relatively wet years ended. According to him, the persistence of below-mean rainfall in the last two decades in Nigeria is an indication of an abrupt change in climate. The agricultural sector in Nigeria is highly sensitive to rainfall patterns, especially in the southern part of Nigeria where rain-fed agriculture is mainly practiced.

Water shortage

Availability of water in Sub-Saharan Africa is highly variable. Sharma *et al.* (1996) reported that eight countries were suffering from water stress or scarcity in 1990 and the situation is getting worse as a consequence of rapid population growth, expanding urbanization, and increased economic development. Around the year 2000, about 300 million Africans lived in a water-scarce environment. It has been predicted that by 2025, the number of countries experiencing water stress will rise to 18 affecting 600 million people (World Bank, 1995). Long-term precipitation records from the Sahara give a clear indication of declining precipitation in that region (UNEP, 1997). These declines in precipitation have reduced hydrological discharges in major river basins in the sub-humid zones of the African continent.

Sensitivity analyses of major rivers of the continent indicate that these rivers are sensitive to climate change. Magadza (1996) examined changes in water storage in Zimbabwe's main water storage facilities during the 1991–1992 drought cycles. During this period, the mean temperature averaged 2°C above the long-term seasonal mean, and seasonal evaporation exceeded the long-term seasonal mean by more than 30%. During this drought cycle, stored water resources dwindled to less than 10% of installed capacity. Urbiztondo (1992) simulated the response of Lake Kariba power generating capacity to various climate change scenarios. His results indicate that with no significant change in precipitation, the lake would regularly fail to meet installed generating capacity if the temperature rose by 4°C. During dry years, generating capacity would decrease by as much as 50%. Even during wet years, maximum generating capacity would barely exceed 50% of installed capacity. With an estimated error of about 8%, the Gambia River flow has been shown to be very sensitive to climate change. Based on the results of river flow responses and vulnerability analysis, climate variables alone can cause a 50% change in runoff in the Gambia River catchments (Jallow *et al.*, 1999). Manneh (1997) reported that generally 1% change in rainfall will result in a 3% change in runoff. This usually results in saltwater intrusion into the Gambia River by 40 km at times of maximum intrusion. The rate of saline water intrusion during dry seasons would increase from about 20 km per month in January to about 40 km per month in April and May.

Coastal erosion and sea level rise

According to the IPCC (1990), there exists a strong correlation exists between greenhouse gas emissions and climate change and between global temperatures and sea level rise. Global temperatures are expected to rise by between 0.2°C to 0.5°C per decade. The rise in temperature is expected to cause thermal expansion of the sea and melting of polar ice. These will cause the sea level to rise for about 3–10 cm per decade during the next century. Another report by the IPCC (2001) revealed that the large scale loss of land ice and thermal expansion of sea water has very likely contributed to the observed sea level rise. According to the International Federation of Red Cross (IFRC, 1999), sea level rise and flooding are already affecting millions of people worldwide. The IFRC report revealed that an estimated 10 million people are at constant risk of coastal floods, and floods in general are making 3 million people homeless every year and that the number of people affected by sea level rise is on the increase annually.

Floods are becoming more frequent in many parts of the world, especially in Africa. Some countries experience floods and droughts at the same time. For example, Mozambique recorded the worst floods in the year 2000 that affected the southern and central regions of the country. In Malawi, when it rains very well in the upper plateau, it causes flooding in the lower plateau. Such phenomena have become very frequent in recent years due to the problem of changing rainfalls coupled with deforestation (Tearfund, 2005).

The Nigerian Environmental Study/Action Team (Okali and Eleri, 2004) reported that sea-level rise and repeated ocean surges will not only worsen the problems of coastal erosion that are already a menace in the Niger Delta of Nigeria, but that the associated inundation will also increase problems of floods, intrusion of sea-water into fresh water sources and ecosystems destroying such stabilizing systems as mangroves, and affecting agriculture, fisheries and general livelihoods. The Niger Delta of Nigeria could lose over 15000 square kilometers of land by the year 2100 with a one meter rise in sea level. Calculations have also shown that a 20 cm rise in sea level will inundate 3400 km^2 of the Nigerian coastland (Onofeghara, 1990). It is estimated that with a sea level rise of 30 cm, about 1 to 2 million people will be affected. In all this, it is predicted that Nigeria will lose about $9 billion as a result of the sea level rise while at least 80% of the people of the Niger Delta will be displaced due to the low level of the region.

Changing seasons and food security

Across Africa, seasons are changing. In Ethiopia, climate is affecting harvest, because a delay in the onset of rains forces farmers to miss the optimum planting time and farmers are unable to plant long-season crops. The growing season has become shorter due to delayed onset and early cessation of rainfall. There is decline in crop yield and poor performance of root crops that require even distribution of rain. In 2004, due to shortage of rain between March and May, a yield loss of 80% was recorded in a region in Ethiopia (Tearfund, 2005).

The rainy season has become unpredictable in Mali. The first rains that used to come in the month of May now fall during the last two weeks of June and the rainy season keeps shrinking. Planting times are affected, because farmers can no longer follow agricultural calendars. The water levels fall year after year and some species of animals and vegetation have disappeared. The survival of the poor people is uncertain (Tearfund, 2005).

Deforestation and Desertification

Forests cover 520 million hectares in Africa, making up 17% of the world's forest. The forest is economically important to the African continent contributing to 6% of GDP. Moreover, about 90% of Africans rely on firewood for their energy requirements. African forests are being destroyed at an alarming rate. In the 1980s, Africa lost 47 million hectares of forest. Climate change will add further stress to this already deteriorating situation. The IPCC predicted that a sustained increase in mean ambient temperature beyond 1°C would cause significant changes in forest and rangeland cover. With global warming forest will be affected by more pest, pathogens and fire. Loss of forests will in turn lead to the encroachment of the desert in Africa.

General Flooding

While climate change will lead to increase aridity and desertification in some parts of Africa, it will lead to flooding in other parts, especially in the coastal regions. Apart from coastal erosion, floods in general have impacted negatively on the livelihoods of many communities in the region. Floods and erosion remove top soil, destroy roads, affect fresh water resources and threaten lives and properties. Many people have been rendered homeless by floods and several roads have been made impassable. The usefulness of several roads has become seasonal, only passable during the dry months of the year. Floods paralyze economic activities in many towns and cities.

Climate change and socio-economic development

Electricity generation in Sub-Saharan Africa is highly dependent on hydroelectric dams. Magadza (1996) reported that in the past half-century, Africa has invested heavily in hydroelectric power schemes. Recent drought episodes highlighted the vulnerability of even the largest hydroelectric plants to climate variability. Due to climate change, hydro-electricity supply will fall, disrupting commercial and industrial activities. The impacts of shortage of electricity can range from inconveniences and discomfort to social deprivation and severe economic loss. To offset the cost of production, many industries sell their goods at higher cost. Some industries may decide to operate below capacity, which may imply cutting down the utilizations of factors of production such as labor.

Climate change will not only affect human societies, but will lead to increases in the spread of pathogens and disease vectors. In many societies, change in availability of water and food could increase health problems due to heat and infections. More important are the consequences of the changes in precipitation and temperature on patterns of disease vectors and viral diseases. Higher levels of ultraviolet radiation due to the depletion of the ozone layer could lead to skin cancer, eye diseases and impairment of body immune systems (Imevbore, 1995). Rapid sea level rise will change coastal ecology and threaten important fisheries; this will have serious impacts on fishery activities, with serious consequences on economies which depend on fishery resources.

Solutions for Sub-Sahara Africa

To effectively solve the problem of climate change will require tackling the menace at local and regional levels. At local level, it is of utmost importance to create awareness among local people. The local people, especially communities that depend on natural sectors need to know the reasons why they are experiencing changes in their environment. Government, policy makers and development partners should make efforts to understand indigenous technical knowledge in the different communities. Indigenous adaptation strategies are peculiar to communities. These peculiarities need to be understood in greater detail; policies and interventions should take into account these peculiarities. In many countries in the world, most developmental efforts are concentrated in the urban areas and little attention is paid to rural communities. Climate change impacts will be most felt in the rural communities in Africa, because they make up the agricultural population and majority which are poor. Governments in Sub-Sahara Africa should give more attention to the development of rural communities. This will help to reduce rural-urban migration and reduce vulnerability to climate change impacts. Particular attention should be given to basic human needs such as access to electricity, water and sanitation.

Many alternatives exist for the supply of water and electricity. The choices available to a society at any given time also depend on factors such as natural resource endowment, technological capability, institutional capacity, financial, market conditions, cultural preferences, awareness, and education. These factors should be carefully studied at the local level to determine alternatives to energy generation and water supply. It is a palpable truism that in combating climate change and poverty, the use of new renewable energy technologies create a decentralized option that generate jobs and income, empowers local communities and strengthens self reliance. Also at the local level, the need to promote energy sovereignty is essential. Renewable energy technologies have the ability to give self reliance to local communities, where they can utilize the excellent renewable energy resources available for their own good. It will allow local people to have control over their energy resources and determine the type of energy to use for their daily needs. In Africa, this issue of energy sovereignty needs to be scaled up and this can only be possible by the use of renewable energy technologies. Renewable energy can also help to limit the insecurity in terms of supply of energy and promote access to more vulnerable populations.

There is, therefore, need to advocate that renewable energy should be integrated into local development plans, while focusing action on the development of rural enterprises (including local production).

At the regional level, the collection, synthesis and analysis of climate data to forecast climatic conditions is imperative in Sub-Sahara Africa. This has been poorly developed in the African region. Monitoring of environmental processes as well as increased weather observations is required. Coordinated collection of integrated data sets for the sub-region in connection with an extreme event such as a drought or flood is highly desirable and would contribute to understanding of adaptation and response strategies and regional integrated modeling. Proper strategies to disseminate climate information to create better agricultural practices in order to minimize losses due to uncertainty are imperative. For Africa to effectively tackle the problems posed by climate change, it is important to train relevant staff on data manipulation, modeling and prediction. Climate change projection will create strong awareness of expected impacts and effects of climate change in order to determine the climate-sensitive sectors in relation to development planning and prioritization for decision-making. It is also relevant for investment negotiations with donors during the preparation of country strategy papers by different donors. Such awareness at the decision-making level will sustain development through country participation in investment prioritization and minimization of climate risk. There is great need for increased African capacity to study the more fundamental science issues of global change and its impacts. There is great capacity at the applied management level in Africa, and this must be strengthened by a strong science capacity. The increasing number of international environmental treaties and agreements will require an even greater capacity for analysis and delivery of timely reports. There also is great need to apply science findings in policy-analysis and international negotiations. The need to train African policy makers and diplomats to increase their capacity to negotiate effectively in international forums is highly desirable.

In conclusion, it is becoming increasingly clear that like most environmental problems, climate change require integration of many disciplines, methods of analysis and strategies. The integration of science findings with management and policy issues are needed. Such integrated models are required at local and regional levels and should include all important linkages between the socio-ecological and economic sectors. Given the unique combinations of factors in Africa (climate, economics, infrastructure), it will be necessary in future to develop and apply regional assessment models and these models should be built around issues of sustainable development rather than emissions reduction.

References

Harmeling, S., Bals, C. and M. Windfuhr, 2007. Climate change and food security in Africa. In: Africa Can Feed Itself. Aksel Naerstad (ed.), AiT AS, Oslo, Norway.

Hulme, M. (ed.), 1996. Climate change in Southern Africa: An exploration of some potential impacts and implications in the SADC Region. Climatic Research Unit, University of East Anglia, Norwich, United Kingdom, 96 pp.

Hulme, M., Doherty, R. M., Ngara, T., New, M. G. and D. Lister, 2001. African climate change: 1900–2100. Climate Research, (in press).

IFRC, 1999. International Federation of Red Cross and Red Crescent Societies (1999). World Disaster Report, 1999. 11 pp.

Imevbore A. M. A., Ofoezie I. E. and E. A. Obot, 1986. Report on the study on snail and cyclopoid borne disease problems of small scale water resources development projects in Nigeria. Submitted to the World Health Organization by the Institute of Ecology, O.A.U., Ile-Ife, Nigeria.

Imevbore, A. M. A., 1995. Some aspects of the impacts of climate change. In: Umolu, J. C. (ed.). Global Climate Change: Impact on Energy Development. DAMTECH Nigeria Limited, Nigeria.

IPCC, 1990. Working Group III, 1990. Strategies for adaptation to sea level rise. Report of the Coastal Zone Management Subgroup, Intergovernmental Panel on Climate Change Working Group, Rikkswaterstatt, The Netherlands. 122 pp.

IPCC, 1992. Global climate change and the rising challenge of the sea. Report of the Coastal Zone Management Subgroup.

IPCC, 2001. Summary for policy makers (A Report of Working Group 1 of the Intergovernmental Panel on Climate Change). Online at www.ipcc.ch.

IPCC, 2007. Climate change 2007: Climate change impacts, adaptation and vulnerability. Summary for policymakers. Genf. http://www.ipcc.ch/SPM13a prO7.pdf.

IPCC, 1998: The regional impacts of climate change: An assessment of vulnerability. Special Report of IPCC Working Group II [Watson, R. T., Zinyowera, M. C. and R. H. Moss (eds.)]. Intergovernmental Panel on Climate Change, Cambridge University Press, Cambridge, United Kingdom and New York, NY, USA, 517 pp.

Jallow, B. P., Toure, S., Barrow, M. M. K. and A. A. Mathieu, 1999: Coastal zone of the Gambia and the Abidjan region in Cote d'Ivoire: sea level rise vulnerability, response strategies and adaptation options. In: National Assessment Results of Climate Change: Impacts and Responses [Mimura, N. (ed.)]. Oldendorf Luhe, Inter-Research, Germany, pp. 129–136.

Lamb, P. J. and R. A. Peppier, 1992. Further case studies of tropical Atlantic surface atmospheric and oceanic patterns associated with Sub-Saharan drought." Journal of Climate, 5, 476–488.

Magadza, C. H. D., 1996. Climate change: Some likely multiple impacts in southern Africa. In: Climate change and world food security [Bowning, T. E. (ed.)]. Springer-Verlag, Dordrecht, The Netherlands, pp. 449–483.

Manneh, A., 1997. Vulnerability of the water resources sector of the Gambia to climate change. In: Republic of The Gambia: Final Report of the Gambia/U.S. Country Study Program Project on Assessment of the Vulnerability of the Major Economic Sectors of the Gambia to the Projected Climate Change. Banjul, The Gambia, (unpublished).

O'Brien, K. and R. Leichenko, 2005. Global environmental change, globalization and food systems. In: IHDP Update 1/2005. http://www.ihpd.uni-bonn.de/html/publications/update/pdf-files/IHDPUpdateFoodl 05.pdf.

Okali, D. and E. O. Eleri, 2004. Climate change and Nigeria: A guide for policy makers. The Publication of the Nigerian Environmental Study Action Team (NEST).

Onofeghara, F. A., 1990. "Nigerian wetlands: An overview." In: Akpata, T.V.I and D. U. U. Okali, (eds.). Nigerian Wetlands, pp. 14–26. Man and the Biosphere (MAB) National Committee, Nigeria, UNESCO National Commission, Federal Ministry of Education.

Riebsame, W. E., Strzepek, K. M., Wescoat Jr., J. L., Perrit, R., Graile, G. L., Jacobs, J., Leichenko, R., Magadza, C., Phien, H., Urbiztondo, B. J., Restrepo, P., Rose, W. R., Saleh, M., Ti, L. H., Tucci, C. and D. Yates, 1995: Complex river basins. In: As Climate Changes, International Impacts and Implications [Strzepek, K. M. and J. B. Smith (eds.)]. Cambridge University Press, Cambridge, United Kingdom and New York, NY, USA, pp. 57–91.

Sharma, N., Damhang, T., Gilgan-Hunt, E., Grey, D., Okaru, V. and D. Rothberg, 1996: African water resources: Challenges and opportunities for sustainable development. World Bank Technical Paper No. 33, African Technical Department Series, The World Bank, Washington DC, USA, 115 pp.

Tearfund, 2005. Dried up, drowned out: Voices from the developing world on a changing climate. Publication of Tearfund, UK.

UNEP, 1997. World Atlas of Desertification, (2nd ed.) United Nations Environment Program, Edward Arnold, London, United Kingdom, 69 pp.

Urbiztondo, R. J., 1992. "Modeling of climate change impacts on the Upper Zambezi River Basin." M. Sc. Thesis, University of Colorado, Boulder, CO, USA, 134 pp.

World Bank, 1995. "Towards environmentally sustainable development in Sub-Saharan Africa: A framework for integrated coastal zone management building blocks for Africa, 2025." Paper 4, Post UNCED Series, Environmentally Sustainable Development Division, Africa Technical Department, World Bank, Washington, DC, USA, 56 pp.

Emmanuel Uzoma Onweremadu
Federal University of Technology, Owerri, Nigeria

Sustaining endangered species in a period of climatic anomaly in Sub-Saharan Africa: Examples from Nigeria

Introduction

There is a symbiosis between the physical resources and the biological life forms within the environment. The biophysical environment, which includes all the variables that comprise the earth's biosphere tend to be in equilibrium without any human activity. But, built environments have resulted in courtesy of anthropogenic activities. Man has reversed nature resulting to loss of resilience in the biosphere. Sub-Saharan Africa was richly endowed with a variety of life forms, residing in aquatic, arboreal and terrestrial homes. Perhaps, sustenance of these original life forms would have weathered atmospheric imbalances as the aforementioned habitats were associated with enormous economic and aesthetic values. These values invariably maintained and supported environmental health. In the age of gathering and hunting, populations in Sub-Saharan Africa were low and localized as there were no international migrations. Growing human populations and increasing mobility in search of livelihoods and other desires exposed Sub-Saharan Africa to new and nature-degrading technologies such as the use of machines in extensive land clearing. Simple traditional technologies of Sub-Saharan Africans were friendlier to its biodiversity. However, there was sudden reversal of value system resulting to attitudinal change. Economic demands increased so rapidly irrespective of negative consequences of actualizing such demands. There was no afterthought of the second side of the coin. Communal life patterns, which used to moderate individual life excesses was subsumed and engulfed by a new life evolving from interactions between Sub-Saharan Africa and the advanced world. This resulted to conflictive natural resource use systems and consequent damage of the Sub-Saharan biotic heritage. In short, demographic pressure and accompanying demands on the environment threatened both the physical and biological resources of the Sub-Saharan biome. These threats are exercised on the biodiversity through economic development, agriculture, urbanization, wars and industrialization. Worst still, some species were traded on without conscious conservation leading to their total extinction or tremendous decline (Table 1). The fact that inhabitants of Sub-Saharan Africa largely depend on natural resources, which are now depleted, points to the high level of poverty in the area. Sub-Saharan Africa is home to more than three quarters of the world's ultra poor, and is the only region in the world in which there are more ultra poor than medial or subjacent poor (Ahmed *et al.*, 2007). The severity of poverty and the limited progress in reducing it indicates that many poor inhabitants may be trapped. Consequently, Sub-Saharan Africa had a Global Hunger Index Score of 25.4 in 2003 (Ahmed *et al.*, 2007), which defines the region's hunger as alarming.

The rate at which life forms are declining prompted the establishment of forest and game reserves. Most animal and plant species live in nature reserves and matrix of managed and human-dominated ecosystems. In these reserves, neither economic development nor resource extraction is allowed (Table 2). This system of setting aside reserves is criticized by scholars (Wilcove, 1989; Hansen *et al.*, 1991), since reserves occupy a small land area in addition to population pressure on such reserves and attendant land use patterns. These inadequacies are becoming more evident with climate change and consequent decline in life forms.

Table 1: Some endangered plant species in Africa

Common Name	Scientific Name	Habit	Use
Monkey fruit	Dactyladenia barteri	Small tree	Forage, fallow crop poles
Oil bean	Pentaclethra macrophyllum	Tree	Fruit/African salad
Alligator pepper	Aframomum melagneta	Herb	Peppery fruit
Elephant palm	Borassus aethiopum	Robust palm	Fruit, palm sap[, log
Rattan	Laccospermia spp Oncochlamus spp.	–	Furniture, basketing
Raffin palm	Raffia hookeri	Tree	Wine, twines, mats, pues
Shea butter	Butyrospermum partkii	Tree	Cooking oil, cosmences chocolate, candy
Bush mango	Irvingia gabonensis	Tree	Condiment, fruit

Common Name	Scientific Name	Habit	Use
Bitter cola	Garcinia cola	Tree	Fruit, medicine
Cola	Cola nitida, Cola auminata	Tree	Stimulant
Gnehim (Ukazi)	Gnetum africanum	Climbing plant	Vegetable
–	prosposis spp.	Tree	Folder
Frankincense	Boswellia papyrifera		Incense, cosmetics medicine, resui
Gum Arabic	Acacia spp.	Tree	Preservative, medicine
Baobab	Adansoma digitata	Tree	Food, fibre, salt (ash)
West African albizia	Albizia adianthifolia	Tree	Shade, medicine
Cocoyam	Xanthosoma sagitifolia	Herb	Food
Breadfruit	Atrocarpus corminis	Tree	Food

Table 2. Diversity and abundance of mesofauna in Akure South Western Nigeria

Forest Habitat	Diversity Index (H)	Evenness (E)	Abundance (M2)
Unlogged forest	1.40	0.10	6100
Lightly logged forest	1.60	0.13	4000
Heavily logged forest	1.19	0.18	2600

Adapted from Nurudeen *et al.*, 2009.

The Problem

Africa has the task of ensuring food security for a population of 1.9 billion people by 2050 and finding jobs for a growing number of rural youth (Spore, 2010). But, about 8 billion tons of carbon is released into the atmosphere, contributing to global warming and climate change (Tewari *et al.*, 2008) (Table 3). Biodiversity decline in Nigeria is obvious and is caused mainly by deforestation for fuel-wood, agriculture, hunting, architectural activities, engineering works for urbanization and industrialization.

Table 3. Emissions of greenhouse gases (GHGs) and aerosols from deforestation and biomass burning.

Species	Emission ratio (x10, 000)	Biomass burning (tg per year)	Allsources (tg per year)	Fraction (%)
Carbon dioxide				
Net from deforestation	2400	7400	32	–
Gross from combustion	880	3590	8790	41
Carbon burned	–	4080	–	–
Carbon monoxide	80	290	1100	26
Methane	8	29	380	8
N_2O	0.1	0.8	6	14
NO_2	2.3	9.6	40	24
CH_3Cl	0.12	1.3	2.3	55
Organic carbon (Aerosol)	25	90	1500	6
Black carbon (Aerosol)	17	59	180	33

Tg = teragram (1 trillion grams)
Adapted from Andreae and Dickson, 1995.

The situation is aggravated by oil exploration and exploitation activities and concomitant spillages (Onweremadu *et al.*, 2008), which they suggested could be ameliorated using composted cassava sludge (Table 4).

Table 4: Effect of composted cassava sludge on removability of contaminants at room temperature with time (days)

Pollutant	Time					
(mg/kg)	30	60	90	120	180	$LSD_{(0.05)}$
Cr	15	25	35	40	50	1.25
Col	35	60	65	75	90	0.92
V	20	25	30	35	45	1.08
Ni	25	40	60	80	85	0.96
TPH	1.8	2.5	6.5	7.5	9.0	0.09

Cr = Chromium, Cd = Cadmium, V = Vanadium, Ni = nickel, TPH = Total petroleum hydrocarbon, LSD = least significant difference
Source: Onweremadu *et al.*, 2008.

It was observed that chromium contained in soils affected by oil spillage significantly (p= 0.05) altered total fungal mycelium, active fungal mycelium, total microbial biomass and respiration (Onweremadu and Nwufo, 2008).Yet, there is only scanty usable data on these ecological problems (Onweremadu *et al.*, 2007a). Manure wastes despite their huge potentials (Onweremadu *et al.*, 2007b) are not utilized to full capacity. Farmers who constitute a greater proportion of land users still stick tenaciously to ineffective traditional land use practices such as bush fallowing and slash-and-burn land clearing. Currently fallow lengths have drastically reduced due to demographic pressure and other conflictive and economically competitive uses (Onweremadu *et al.*, 2007c), consequently the biotic environment is threatened despite the fact that a great majority of domesticated and wild plants are good sources of nutrients (Ogoke *et al.*, 2009) (Table 5).

Table 5: Some nutritional composition (g/100g) of selected grain legumes in the cropping systems of Southeastern Nigeria

Legume	Carbohydrate	Protein	Oil	Fibre
Cowpea	1	22.5	1.4	5.4
Pigeon pea	65	19.5	1.3	7.3
Lima bean	61	21.4	1.4	4.3
Soybean	33	33.7	17.9	4.7
Bambara groundnut	61	18.8	6.2	4.8
African yam bean	67	19.2	1.1	5.2
Groundnut	23	23.2	11.8	2.9

Adapted from Ogoke *et al.* 2009.

There is a strong linkage between temperature and plants (Weber and Stocks, 1998). Obioh (2002) noted that forest ecosystems are vulnerable to temperature abnormality, causing instability and disturbances at the landscape level (Turner, 1998). Forest disturbances such as fires, drought, convective storms, insect and pathogen outbreaks, sandstorms, erosion and landslides affect life forms. Changes in structure, functional processes and composition are primed by forest fires (Justice and Townsend, 1981), leading to decline in value (Njoku, 2009). With pronounced change in decadal mean temperature from 1970–2005 (Table 6) in Southeastern Nigeria (Njoku, 2009), it is obvious and implied that alterations have occurred in the species composition and population. Indeed, a study by Njoku (2009) showed imminent shortage of forest resources, products and services arising from temperature anomaly.

Table 6: Climatic data for 5-year monthly average (1971–1975) for (Nsukka, Latitude $6^0 52^I$N, Longitude $7^0 24^I$E, Altitude 400m)

Month	Rainfall (mm)	Rain days (No.)	Sunshine (1hrs)	Solar radiation (Cal/cm2 /day	Temp. (Max.)	Temp. Min.	Vapour pressure (mb)	Wind speed (m/ sec.)	Relative humidity (10 hrs.)	Relative humidity (16hrs.)
Jan.	5.9	1	5.1	69.2	31	20	10.5	1.32	56	37
Feb.	14.9	1	6.7	289.3	33	22	20.3	1.52	69	43
Mar.	37.0	4	5.5	285.7	32	23	24.2	1.72	73	51
Apr.	121.2	8	6.1	285.7	31	22	25.8	1.58	76	62
May	216.8	11	5.7	272.9	30	21	27.8	1.30	79	68
June	241.9	14	3.9	243.6	28	21	25.8	1.32	83	75
July	148.4	15	3.3	214.3	27	21	26.2	1.36	87	78
Aug.	197.7	16	3.8	208.8	27	21	24.6	1.42	88	77
Sept.	372.5	20	3.7	232.6	27	20	24.4	1.18	86	76
Oct.	205.1	14	5.2	250.9	28	21	25.0	1.05	82	72
Nov.	8.0	2	6.3	300.3	31	20	24.5	1.09	68	53
Dec	17.7	1	4.8	285.7	31	20	16.0	1.27	53	41

Adopted from Ezedinma, 1976.

Rainfall is a climatic attribute whose variability will obviously affect biodiversity performance. Rainfall varies unpredictably in amount, intensity and distribution. These rainfall characteristics co-vary in permutations to produce phenomenal drought even during conventional wet seasons. Deviations in these rainfall characteristics, which are more pronounced under climate change, can lead to droughty conditions, resulting to biodiversity declines or total extinction due to low resilience during such aberrations (Table 7).

Extreme inter-annual variability of precipitations is common in Nigeria, and indeed other Sub-Saharan African countries, thereby inducing either droughts or floods. This creates untold repercussions on biodiversity including agricultural productivity.

Table 7: Rainfall and temperature data for 2006/2007 in Owerri, South-eastern Nigeria (Mean monthly data)

	– Rainfall (mm) –		–Temperature (°C) –	
Months	2006	2007	2006	2007
January	89.8	0.0	26.0–31.0	26.0–31.5
February	1.8	46.8	27.6–31.7	26.5–31.0
March	167.0	78.0	24.0–31.2	29.5–31.2
April	81.9	135.2	25.0–31.1	23.0–31.5
May	358.2	393.3	23.0–31.5	24.0–32.0
June	454.7	407.6	22.5–30.0	24.5–31.5
July	625.0	311.2	24.8–28.3	24.4–28.5
August	286.7	246.3	24.0–28.0	23.8–28.5
September	479.5	472.7	22.1–28.0	24.5–29.2
October	360.6	165.5	23.0–29.5	24.0–29.5
November	9.2	85.8	26.2–31.0	24.1–30.2
December	0.0	19.7	26.5–31.0	26.1–31.8
Total	2914.40	2362.10	–	

Source: Meteorological Department, Federal Ministry of Aviation, Owerri, Nigeria.

Long duration and high intensity rainfalls and water erosion damaged landscapes are recognized as inter-rill, rill, rain splash and gully. Gullies have devastated landscape integrity, destroying life forms in southeastern Nigeria (Onweremadu, 2007) due to dispersion and disruption of water-unstable aggregates in the area (Onweremadu, 2008). Soils are lost and build-up sediments in natural water bodies and valleys forcing water to spill over adjoining terrestrial habitats as floods (Table 8). In addition to this, sediments pollute aquatic habitats and inhabitants resulting in aquatic biodiversity decline. Flood events are now characterizing Northern Nigerian landscapes due to sporadic heavy rainfall; and the runoff there-from cannot be contained in ephemeral water courses (Marigots). This is aggravated by collapse of dams and irrigation structures. Some of these floods hang for a longer time on soil bodies, creating anaerobic conditions, and consequent death of aerobic life forms. Resulting floods after abundant rainfall in the Sahel savannah subside in two to four days if no other rain occurs in the meantime (Tauer and Humborg, 1992).

Table 8. Obstacles to adoption of soil information/ data (N = 450)

	Factor 1	Factor 2
Variable	Management/Economic constraint	Technical and structural constraint
Language of delivery	0.78	0.09
Shortage of trained personnel	0.18	-0.86
Lack of usable information	0.36	0.69
Scanty soil data	0.88	-0.52
Poverty	0.68	0.26
Land tenure	0.16	0.66

Adapted from Onweremadu *et al.*, 2007c.

Wind factor of climate is known to erode top soils, which are deposited on life forms on natural and artificial waterways, leading to unmanaged and catastrophic floods. It is also true that amount of precipitation depends on global winds, which is influenced by the periodic dislocation of the inner-tropical conveyance zone (ITC). Dry winds increase evapotranspiration, which limits moisture available to plants, causing them to display xerophytic, geophytic and therophytic behavior in response to harsh climate. A majority of these life forms become vulnerable when critical limits in climatic variability are exceeded.

Another major factor associated with biodiversity decline is the rising ocean levels. Coastal erosion is now a threat to inhabitants as the intrusion of salt water beyond the shoreline is invariably increasing soil salinity. The salinization of coastal soils increases exchangeable sodium percentage, which in turn causes disaggregation of soil aggregates (peds) thereby heightening erodibility of soils. Un-eroded patches on the coast become so saline that only halophytes can thrive with favorable performance. Extinction of life forms on saline coastlands would certainty lead to the formation of so many marine deserts at the coasts of world oceans including the Atlantic Ocean.

Disappearing forests and other vegetal forms are a great threat to animal life. Loss of forests and forest resources implies destruction of sun-hating plants, saprotrophs/detritivores, creeping and climbing forms, reptiles, arboreal organisms, etc. In addition to this, deforestation promotes low soil carbon sequestration, increasing volatilization, increasing thermal impact and severe land, water and air degradation. Agricultural development, logging and ranching are often identified as the proximate causes of deforestation (Cattaneo, 2002). Deforestation reduces canopy cover, disturbs and impoverishes forest floor, and this can lead to soil erosion and loss of forest function, especially in terms of hydrology and biodiversity (Liniger and Critchley, 2007).

While microorganisms play crucial roles in regulating ecosystem processes such as nutrient mineralization, soil carbon storage, trace gas fluxes, transformation of aqueous solutes and processing of water pollutants, deforestation disrupts these functions. Interactions among plants, soil, hydrology and microorganisms regulate nutrient cycling in ecosystems (Barthuiya *et al.*, 2008) and these interactions vary in time and space (Vitousek *et al.*, 1994) but affect aboveground plant productivity (Zak *et al.*, 1994), suggesting their dependence on inputs of reduced carbon in the soil through litter (Allen and Schlesinger, 2004) (Table 9).

Table 9: Effect of trivalent chromium on soil microbes

Rate	Total fungal mycelium	Active fungal mycelia	Total microbial biomass	Respiration
Mg Cr/kg	(mg/g d.w.)	(mg/g d.w.)	Carbon (mg/g d.w.)	CO_2 (Mg/gd.w./hr)
0	0.25+0.06	0.13+0.01	0.61	41
10	0.18+0.07	0.09+0.01	0.58	40
50	0.17+0.01	0.07+0.03	0.65	39
100	0.15+0.03	0.06+0.02	0.60	40
250	0.16+0.02	0.04+0.01	0.61	38
500	0.13+0.02	0.03+0.01	0.62	37
1000	<0.001	<0.001	<0.001	<0.001

d.w. = dry weight, Cr = chromium
Adapted from Onweremadu and Nwufo, 2008.

Logging is becoming inevitable due to its importance in the Power Holding Company of Nigeria for electrical transmission and other uses (Njoku and Akachukwu, 2009). The number of logs harvested in the tropics is not necessarily the primary threat to survival of tropical biodiversity but the destructive manner in which logs are harvested (Muul, 1993), leading to negative impacts on diversity and abundance of mesofauna in rainforest vegetation of Western Nigeria (Nurudeen *et al.*, 2009). Logging exposes soils to intense sunlight which drastically reduces ant populations (Badejo *et al.*, 2005) and other soil organisms (Crossley and Blumberg, 1983). This affirms that land use alters physicochemical and biological characteristics of soils and ecosystems.

De-vegetation, climate change and biodiversity

There is a vicious cycle in which human activities such as burning of fossil fuels, and deforestation emits 6 billion tons of CO and 1.6 billion tons of CO_2, respectively into the atmosphere. Ruminating cattle emit 100 million tons of methane into the atmosphere yearly. Termites produce 15 to 35 million tons of methane annually by fermenting vegetable matter from tropical forests in their intestine. Di-nitrogen oxide or nitrous oxide is emitted during intensive agriculture activities involving pesticide use, deforestation and chemical processes including production, formation and destruction of aerosols. There exist relationships between sources of atmospheric trace gases, aerosols and environmental consequences, as the chemistry of the atmosphere has evolved with the biota (Andreae and Dickinson, 1995). Drivers of atmospheric change such as agriculture, animal husbandry, fisheries, biomass burning, fossil fuel combustion and industrial processes interact with their effects on the environment, namely ultraviolet radiation, global warming, photochemical smog, global oxidation efficiency, acid rain, visibility and corrosion to alter biotic systems.

Conserving endangered biota

Biodiversity protective measures should create an arrangement of systems in space, which influences the intrinsic stability of ecosystems and the ability of ecosystems to provide a variety of human subsidies, including food, fiber, water and recreation. Focus should be on connectivity between patches of natural systems. It is believed that the longer the ecotone, the more suitable it is for wildlife. Ecotones uniquely harbor rare and endangered species.

Climate change effect is location-specific, implying that conservation measures should be in tandem with the bio-geographical requirements of a place. Considerations for a stable biodiversity include an understanding of Justus Liebig's Law of the Minimum and Shelford's 1911 Law of Tolerance (Law of Maximal Limits). While the former emphasizes on the least available essential growth factors as the most determining factor for optimal productivity, the latter states that excess of a critical maximum may exclude certain organisms from a landscape or area. There is therefore the need to strike a balance between environmental factor inadequacies and surpluses. A good understanding of pedological, zoological, hydrological, climate-logical, anthropological, psycho-logical and economic interactions become paramount and are necessary in recovering and sustaining the biosphere.

Knowledge of these areas of knowledge will help in protecting biodiversity. A knowledge-driven approach is necessary particularly in Africa where women bear an average of 5.5 children, with the continent having the fastest-growing and most youthful population in the world (Spore, 2010). Population increase implies heightened pressure on biodiversity. Based on the above, the following protective and restorative approaches are hereby suggested for endangered species.

Organized cultivation of rare, endangered but valuable plant species can recover and sustain the disappearing wild biotic resources. It is important to recall that all agricultural crops and animals were wild plants and animals before their cultivation and rearing based on attributes of popularity and marketability. Wild endangered plants can be conserved and domesticated *ex situ*. Alternatively, those wild species that are difficult to establish when transplanted should be conserved in their natural homes, and this is referred to as *in situ* conservation (See Tables 10 and 11).

Table 10. Some common tree legumes in the cropping systems of Southeastern Nigeria

Scientific Name	Common name	Local name	Parts used
Brachystegia nigerica	–	Achi	Seed
Diallim guineense	Velvet tamarind	Icheku	Seed
Tetrapleura tetraptera	Aidon tree	Ushakrisha	Fruit
Parkia clappertoniana	Locus bean tree	Okwuma	Fruit, leaves, bark

Adapted from Ogoke *et al.*, 2009.

Table 11. Maize performance using animal manures

Animal manure	Plant height (cm) at 6WAP	DM yield (g/pot)
Muturu dung	82.9	348.0
Goat dung	83.5	3242
Sheep dung	83.7	326.3
Pig dung	83.9	355.8
Control	38.4	110.6
$LSD_{(0.05)}$	9.2	6.4

WAP = Weeks after planting, DM = dry matter yield; LSD = Least significant difference.

Adapted from Onweremadu *et al.*, 2007b.

Conservation tillage techniques are advocated for minimal disturbance of life forms. Conservation agriculture operates on three principles, which aim to disturb the soil minimally, cover soil as much as possible while mixing and rotating crops such that the soil and soil-inhabitants are at no time exposed to harsh and excruciating climate. Conservation agriculture lays the stover on the soil, controls grazing and stocking, opens narrow planting furrow with a ripper or makes small planting holes, uses compost, green manure, farmyard manure, legumes and supplemented with appropriate inorganic fertilizer. Conservation agriculture uses mulch and cover crops to suppress weeds, uses fences to confine livestock and establishes fodder banks for farm animals. Conservation agriculture fits well with two of the major NEPAD (New Partnership for Africa's Development) sectoral priorities: Agriculture and environment, and these contribute significantly to the Millennium Development Goals No. 1 (Eradicate extreme poverty and hunger), No. 2 (Ensure environmental sustainability) and No. 3 (Develop a global partnership for development), (IIRR & Act, 2005).

Tree-based improved fallow systems and the biomass transfer systems are helpful in areas characterized by low inputs and low crop productivity (Place *et al.*, 2005). Improved fallows involve intentional use of high performing fallow plants to regenerate or replenish soil nutrients in a short time. Within the farming system, woody plants protect the environment, act as companion crops; and act as shade to the livestock. Again, woody plants produce green manure; fodder, fuel-wood, stakes for farm use and yield fruits, fiber, timber, ornamentals, etc. They also provide homes for arboreal organisms. Agro-forestry reclaims impoverished African soils (Sanchez *et al.*, 1997).

Verheij (2003) remarked that trees disappear in the tropics because more care is given to trees that are owned and grown due to high value, while spontaneously growing ones (wild trees) are neglected. These neglected trees are highly endangered. A combination of traditional (evolved through experience accumulated by generation of farmers) such as home gardens, parkland trees and live fences; modern (evolved with the help of agricultural science) such as shelterbelts, kind hedgerow barriers, or recent (evolved with the help of research in agro-forestry) such as alley cropping and improved fallows, will be effective in sustaining the biodiversity for future generations. Some of these technologies can be adopted by individuals or communities and practiced for a long time for them to achieve efficacy. Community nurseries can be established and seedlings distributed to farmers. If all farmers transplant seedlings and grow them up to maturity, it will have a positive and cumulative effect on the environment (Table 10).

Organic farming builds a pedosphere (soil environment) replete with micro-, meso- and macro-flora and -fauna. Organic farming promotes organic matter build-up in soils via principles aimed at reducing tillage, diversifying cropping systems, eliminating use of herbicides, applying manure, using crop rotation, cover crops, green manures, night soil, mulches, rehabilitating degraded lands and planting/replanting trees (buffer or contour strips, agro forestry, etc.). These practices create a living soil. One gram of good soil contains millions of organisms and vast biodiversity (LEISA, 2008). Soil food web consists of an interaction between soil organic matter, plant roots, arthropods (shredders, predators), nematodes (root feeders predators, fungi and bacteria feeders), fungi (saprophyte), mycorrhizal fungi, protozoa (amoeba, ciliates, flagellates), viruses (bacteriophages, fungi ores), birds and other animals. The richness of soil biodiversity influences the status of aboveground biodiversity, especially in the terrestrial ecosystems (Tables 11 and 12).

Table 12. Characterization of livestock droppings

Property	Unit	Cattle	Poultry	Pig
pH water	–	8.5	7.8	8.0
OC	g/kg	136.0	195.0	156.0
TN	g/kg	5.0	24.6	11.2
C/N	–	25.0	8.0	13.9
CEC	Cmol/kg	1.8	6.6	3.2
Ca	Cmol/kg	0.5	1.2	0.7
Ca/Mg	–	3.6	5.5	4.6
Available P	Mg/kg	1.1	3.8	2.2
Pb	Mg/kg	4.8	10.6	8.7

OC = organic carbon, TN = Total nitrogen, CEC = cation exchange capacity, Ca = Calcium, Mg = Magnesium, P = phosphorus, Pb = lead.

Adapted from Onweremadu *et al.*, 2010.

An integrated approach to climate change adaptation is a panacea to decline in biodiversity (Gurung and Bhandari, 2008). Climate affects all aspects of human endeavors, implying the need to integrate ecosystem activities and socioeconomic ventures in shaping local livelihoods. An integrated approach as it affects biodiversity looks at agricultural production, water resources management, forest, land an soil conservation, diversification of incomes and livelihoods, institutional development, reconstruction of infrastructure as well as awareness and education as essential considerations in adapting to climate change for protecting life forms. Efficacy will be achieved if traditional and modern conservation practices are blended. In agriculture for instance, traditional practices such as multiple cropping, mulching, green manuring, etc., could be blended with vermin-composting, use of improved seeds, seed banks and crop diversification. Breeding climate-proofing life forms is suggested and could be done through establishment of community botanic gardens, where experts and stakeholders come together to share ideas. Breeding should consider evolving salt-tolerant plants that can withstand the salinization of coastland soils. Desalinization of salt water is practicable but expensive and unaffordable to a good number of natural resource users at community level.

Conserving biodiversity appreciably can be promoted by training and re-orientating natural resource users. This can be achieved through formal and informal education in schools, village gatherings, churches and mosques. Schools, religious bodies and community festivals attract large crowds, which provide the audience for this purpose. The Extension Service encourages and manages demonstrations, which are organized on-site to promote participatory approaches to protecting biodiversity. Natural resource users should be organized into climate field schools (CFS). The CFS should be a participatory extension approach in which natural resource users are taught how to manage life forms around them. Nature entrepreneurs are advised to design environmentally-friendly equipment and products.

Policy support is necessary. Recently, a state legislative arm criminalized bush burning in Southwestern Nigeria. Government policies should aim at safeguarding biodiversity and securing food for man. Eco-agriculture is a new and desirable paradigm. Dynamically conservation systems should include preserving traditional systems, which promote minimum tillage. Soil and biodiversity conservation in older farming systems is poorly acknowledged but they are less "biocidal". Policy support should focus on appreciating land users' eco-friendly efforts and inputs. Subsidies and incentives catalyze spread of conservation practices, especially new and innovative ones. Government should promote research on biodiversity conservation while upgrading curricula in all disciplines in present realities. Government assistance is needed in publicizing adaptable plant species such as Moringa and Jatropha for their multi-purpose uses. Again, there is need for an auspicious environment allowing collaboration between government and non-governmental organizations. Laws should be enacted to secure tenure rights for these non-governmental bodies. Finally, policies should be tailored towards encouraging competitive participation in nature conservation, and this can be garnished with green awards to those who have exemplified themselves in protecting biodiversity amidst a changing atmosphere.

References

Ahmed, A. U., Hill, R. V., Smith, L. C., Wiesmann, D. M. and T. Frankenberger, 2007. The world's most deprived characteristics and causes of extreme poverty and hunger. Int. Food Policy Res. Inst., Discussion paper 43, Washington D. C., 130 pp.

Allen, A. and W. Schlesinger, 2004. Nutrient limitations to microbial biomass and activity in loblolly pine forests. Soil Biol. Biochem., 36: 581–589.

Andreae, M. O. and R. E. Dickinson, 1995. Sustainability and changing atmosphere: Assessing changes in chemical and physical climate. In: Munasinghe, M. and W. Shearer (eds.): Defining and measuring sustainability: The biophysical foundations. The United Nations University UNU) and the World Bank, Washington DC, USA. Pp. 117–134.

Badejo, M. A., Tian, G. and I. O. Okoh, 2005. Litter transformer as ecosystem engineers. Sust. Agroecosyst., 2: 91–94.

Barbhuiya, A. R., Arunachalam, A., Pandey, H. N., Arunachalam, K. and M. I. Khan, 2008. Effects of anthrogenic disturbance on soil microbial biomass C. N. and P. in a tropical rainforest ecosystem of Assam, Northeast India. Malay. J. Soil Sci., 12: 31–44.

Cattaneo, A., 2002. Balancing agricultural development and deforestation in the Brazilian Amazon. Int. Food Policy Res. Inst., Washington D C, Research Report, 129. 146 pp.

Crossley, D. A. and A. Y. Blumberg, 1983. Comparison of soil surface arthropod populations in conventional tillage, no tillage and old field systems. Agroecosyst., 8: 249–253.

Ezedinma, F. O. C., 1976. Weather and annual crop production in the rainforest zone of Southern Nigerian. In: Anazodo G. U. and E. U. Odigboh (eds.), Appropriate approaches for accelerated food production in the rainforest zones of Nigeria. University of Nigeria, Nsukka, Nigeria. pp. 59–66.

Gurung, C. B. and D. Bhandari, 2008. An integrated approach to climate change adaptation. LEISA, 24 (2): 6–8.

Hansen, A. J., Spies, T. A., Swanson, F. J. and J. L. Ohmann, 1991. Conserving biodiversity in managed forest: Lessons from natural forest. Bioscience, 41: 382–392.

IIRR & ACT (International Institute of Rural Reconstruction & African Conservation Tillage Network), 2005. Conservation agriculture: A manual for farmers and extension workers in Africa. Majestic Printing Works Ltd., Kenya, 251pp.

Justice, C. O. and J. R. G. Townsend, 1981. The use of landsat data for land cover inventories of Mediterranean lands. In: Townshend, J. R. G. (ed.), Terrain analysis and remote sensing. George Allen and Urwin, London. LEISA (Low External Input and Sustainable Agriculture), 2008 Living soils LEISA 24(2): 4–5.

Liniger, H. and W. Critchley (eds.), 2007. Where the land is greener: Case study of soil and water conservation initiatives worldwide. WOCAT, CTA, FAO, UNEP and CDE. Stampfil AG., Bern. 364 pp.

Muul, I. 1993. Tropical forests, integrated conservation strategies and the concept of critical mass and biosphere (MAB). Digest No 15, UNESCO, France.

Njoku, J. D., 2009. Temperature anomaly and responses by rainforest cover: Potential implications for food availability in Southeastern Nigeria. Proceedings of the International Conference on Global Food Crisis held at Owerri, Nigeria, April 19–24, 2009. Pp. 281–287.

Njoku, K. U. and A. E. Akachukwu, 2009. Variation in deterioration of untreated teak (*Tectonia grandis*) poles inoculated with white rot fungus (*Trametes cingulata*) and brown rot fungus (*Lentinus lepideus*). Int. J. Agric. Rural Dev., 12(1): 11–18.

Nurudeen, T. A., Adebola, S. A., Onyeanusi, A. E., Ekpo, E. N. and V. O. Dania, 2009. Post impact assessment of logging on mesofauna and soil properties. Int. J. Soil Sci., 1(1): 1–7.

Obioh, I. B., 2002. Evidence of climate change and why it is a concern for Nigeria. Paper presented at the Canada Nigeria climate change capacity development. Project NEST – GCST Awareness Raising Workshops, Ibadan Center for Energy Research and Development, Obafemi Awolowo University, Ile-Ife.

Ogoke, I. J., Ibeawuchi, I. I., Ngwuta, A. A., Tom, C. T. and E. U. Onweremadu, 2009. Legumes in cropping systems of Southeastern Nigeria. Jour. Sust. Agric., 53:823–834.

Onweremadu, E. U., 2007. Pedology of near gully sites and its implications on the erodibility of soils in central southeastern Nigeria, Res. J. Environ. Sci., (2): 71–76.

Onweremadu, E. U., Asiabaka, C. C., Adesope, O. M. and N. S. Oguzor, 2007a. Application of indigenous knowledge on land use activities among farmers in Central Southeastern Nigeria. Online J. Earth Sci., 1 (1): 47–50.

Onweremadu, E. U., Asiabaka, C.C., Anyanwu, C. P., Anyanwu, G. A., Etuk, E. B. and B. O. Esonu, 2007b. Survey of animal manure production, management, use and effect on degraded Kandiudult. Int. J. Agric. Res., 2(11): 929–936.

Onweremadu, E. U., Akamigbo, F. O. R., Igwe, C. A., Matthews-Njoku, E. C. and P. C. Obasi, 2007c. Constraints to adoption of soil survey information by owner-managed arable farmers of the humid tropics. Int. J. Agric. Res., 2(8): 712–718.

Onweremadu, E. U., Ofurie, O. E. and S. U. Onwudike, 2010. Soil properties and led concentration in livestock based land use types. Nig. J. Soil Sci., 20(2): 63–63.

Onweremadu, E. U., Ndukwe, O. C. and G. F. Okorafor, 2008. Characterization and removability of priority pollutants in an oil-spilled site using composted cassava sludge. Life Science J., 5(3): 62–66.

Onweremadu, E. U. and M. I. Nwufo, 2008. Pedogenetic activities of soil microbes as influenced by trivalent cationic chromium. Res. J. Soil Viol., 1(1): 8–14.

Onweremadu, E. U., 2008. Evaluating soil structure and hydraulic conductivity by land use in Nigeria. Soil Survey. Horizon., 49: 6–11.

Place, F., Adato, M., Hebinck, P. and M. Omosa, 2005. The impact of agro forestry based soil fertility replenishment practices on the poor in Western Kenya. Int. Food Policy Res. Inst., Washington DC, Research Report, 142. 166 pp.

Sanchez, P. A., Shepherd, K. D., Soule, M. J., Place, F. M., Buresh, R. J. and N. Izac, 1997. Soil fertility replenishment in Africa: an investment in natural resource capital. Soil Soc. Am. Special Publ., No. 51: 46 pp.

Spore, 2010. Rural challenges 2010–2050: An uphill struggle, population and agriculture. The CTA, Spore Special Issue, pp. 7–9.

Tauer, W. and G. Humborg, 1992. Runoff irrigation in the Sahel zone: Remote sensing and geographic information systems for determining potential sites. CTA, Wageningen, The Netherlands. 192 pp.

Tewari, A., Singh, V. and P. Phartiyal, 2008. The potential of community managed forests for carbon trade. LEISA, 24 (4): 32–33.

Turner, M. G., 1998. Factors affecting succession: Lessons from large infrequent natural disturbances. Ecosystems, 1:511–523.

Verheij, E., 2003. Agro forestry. Agrodok. Series, No. 16, Agromisa Foundation, Wageningen, The Netherlands, 87 pp.

Vitousek, P. M., Turner, D. R., Parton, W. J. and R. L. Sanford, 1994. Litter decomposition on the Muna Loa environment matrix, Hawaii: Patterns, mechanics and models. Ecol., 75: 418–429.

Weber, M. G. and B. J. Stocks, 1998. Poorest fires and sustainability in the Boreal forests in Canada, AMBIO, 27: 545–550.

Wilcove, D. S. 1989. Protecting biodiversity in multiple use lands: Lessons from the U.S. Forest Service: Trends in Ecology and Evolution, 4: 385–388.

Zak, D. R., Tilaman, D., Parmenter, R., Rice, C. N., Fisher, F. M., Vose, J., Milchuchas, D. and W. Martin, 1994. Plant production and soil microorganisms in late successional-ecosystems: A conditional scale study. Ecol., 75: 2333–2347.

III. Actions implemented in facing climate change

A.O. Adeyina, O.A. Adeyina and T. R. Feyeye
University of Ilorin, Nigeria

Protection of animal species endangered by climate change in Sub-Saharan Africa

Abstract

Global warming is the single most urgent threat to the future of wild-life and wild-life habitats in Sub-Saharan Africa (SSA). Of the 28 countries reported worldwide to be exposed to extreme risks of global warming, 22 are in Sub-Saharan Africa. Within this region of Africa, animal species such as the Dama gazelle, Umbrella thorn, Scimitar-horned oryx, Spotted hyena, Cameroon Clawless Otter, Demoiselle crane, Mandrill, Potto, the Gray parrot, Bongo, and Clarke's Gazelle are currently facing varying degrees of endangerment. The impacts of global warming pose two major problems across the sub-continent: changes in phenology (timing of season elements) and changes in distribution of wildlife. Low availability of forage resulting from drought have led to habitat loss and overcrowding around few lakes, eliminating in the outbreak of diseases terrestrial species like the Grevyzebra. Sharks, Dolphins and Whales are increasingly threatened by climate changes that are altering ocean circulation, sea surface temperatures, and even the chemistry and salinity of the ocean. Preservation of endangered animal species of SSA is important, not only on economic grounds, but also because they offer a vast genetic "library" from which we can withdraw many useful items in the future. A first priority should be a comprehensive compilation of native animal species in SSA countries and documentation of the status of each of them. This should be followed up with practical habitat protection actions. The creation of a professional society dedicated to ecological ethics to help in research and management of biodiversity is also a good action. Finally, promotion of management actions that are also backed by legal mechanisms through the promulgation of an Endangered Species Act shall go a long towards protection of animal species endangered by climate change in Sub-Saharan Africa.

Introduction

The fragile geographical position of most countries in the Sub-Saharan sub-continent of Africa has made the sub-region one of the most vulnerable regions to the debilitating effects of climate change. The attention of scientists and policy makers in different parts of the world has been drawn in recent years to the impact and vulnerability of livestock production under climate change. An essential but often neglected component is the impact of climate change on the undomesticated species, which form the reservoir of untapped genetic materials for meeting the ever increasing demand for animal protein in recent years. Kostyack and Rohlf (2008) observed that global warming is the single most urgent threat to the future of wild-life and wild-life habitat. Information on the threat of climate change to native undomesticated animal species and the means to ameliorating such effects is essential to preserving the endangered species from total extinction.

Global warming refers to a measurable increase in the average temperature of the earth's atmosphere, oceans and land masses which in turn causes changes in seasons, wind velocity and the amount of precipitation and humidity. Scientists believe that the earth is currently facing a period of rapid warming brought about by rising levels of heat trapping gases known as greenhouse gases(GHGs), in the atmosphere. Greenhouse gases occur naturally, and without them the planet would be too cold to sustain life. However, the beginning of the industrial revolution in the mid 1700s increased human activities, which have added more of these gases into the atmosphere. The Intergovernmental Panel on Climate Change (IPCC, 2007) concluded that human activity is "very likely" causing the world to warm. These reports also noted that the average surface temperature of the earth increased by 0.76 degrees Celsius (°C) in the 20th century and predict that this temperature will likely increase by another 1.8 to 4.0°C in the 21st century, depending on pollution levels. The same IPCC (2007) report also stated that 20–30% of species worldwide are likely to be at increased risk of extinction if increases in average global temperatures exceed 2 to 3°C above pre-industrial levels. The Third Assessment Report of the UNEP/WMO Intergovernmental Panel on Climate Change (IPCC) noted a warming of approximately 0.7°C over most of the African continent during the 20th century (UNEP/GRIDA, 2008).

Endangered Animal Species in Sub-Saharan African

Nature relies on a dedicated balance of plant and animal life to remain healthy. Although change is a natural part of the evolving world, changes that occur at a pace that is too rapid usually have disastrous effects on individual species of animals and plants. Under the International Union for Conservation of Nature (IUCN) Categories and Criteria, *endangered species exist* between critically endangered and vulnerable levels (Bramov *et al.*, 2009). Examples of the various IUCN categories include:

- *Extinct:* All individuals of the group are dead.
- *Extinct in the wild*: Captive individuals survive, but there is no natural population.
- *Critically endangered*: Faces an extremely high risk of extinction in the immediate future.
- *Endangered*: African wild dog faces a very high risk of *extinction in the near future.*
- *Vulnerable*: Cheetah and lion face a high risk of extinction in the medium-term.
- *Conservation dependent*: Spotted Hyena is not severely threatened.
- *Near threatened*: Solitary eagle may be considered threatened in the near future.
- *Least concern*: No immediate threat to the survival of the species House mouse.

A recent risk assessment revealed that out of the 28 countries exposed to extreme risk, 22 are in Sub-Saharan Africa thus endangering the valued animal species. The endangered animal species include the Dama gazelle, Umbrella thorn, Schimitar-horned oryx, spotted Hyena, Cameroon clawless otter, Demoiselle crane, Mandrill, Potto, Gray parrot, Bongo, Southern double-collared sunbird, False Mufuti, Southern ground hornbill, Sable antelope, Bontebok, Aye-aye, Clarke's gazelle, Coquerel's sifaka, Damagazelle, Dugong, Malagasy Giant Jumping Rat, Red-eared Nose-spotted Monkey, Sand cat, Western Giant Eland, White-throated guenon, Addax, Ostrich, Crocodiles, Cape mole rat and Geometric tortoise. Other examples are seen in Table 1.

Table 1: Some endangered animal species.

Name	Zoological Name	Habitat	Description	Present population	Threat	IUCN listing
African dog	Lycaon pictus	Savannas grass-lands and open woodlands are preferred habitats.				
Cheetahs	Acionyx Jubatus	Open and partially open savannah of eastern, central and south- western Africa.	0.8–0.9 m at the shoulder 50–64 kg in weight, 113 km/hr in top speed, with a life-span of 10–12 years.	9.000 to 12.000 cheetahs remain in the wild in Africa.	Habitat loss, conflict with humans, predation and inbreeding due to isolated populations.	Vulnerable
Bats	Order Chirop-tera suborders: Megachirop-tera and Micro-chiroptera.	In caves, crevices, tree activities and buildings. Some species are solitary while others form colonies.	Largest bats have a 6 foot wingspan. Brown bat can live for about 40 years.	Population of some bats are dangerously low or in decline.	Habitat destruction, wind turbines.	Some species are endangered or threate ned (ESA listing.

Name	Zoological Name	Habitat	Description	Present population	Threat	IUCN listing
Chimpanzees	Pantroglodytes	Prefer dense tropical rainforests but can also be found in secondary -growth forests, woodlands, bamboo forests, swamps and even open savannah.	Approximately 4 feet high, males weigh between 90 and 120 pounds, while females weigh between 60 and 110 pounds with a lifespan of 50 to 60 years.	100.000 to 200.000 chimpanzees remain in the wild.	Habitat destruction is the greatest threat, hunting and com-mercial exportation.	Endangered
African Elephant	Loxodonta	African savannah Elephants are found In savanna zones South of the Sahara Desert. African forest west and central Africa.	5–14 ft at shoulder for males. Length: up to 30 feet trunk to tail. Weight: 6.000–15.000 lbs (males) and with a lifespan is 40 years.	450.000–700.000 African elephants.	Habitat loss, pouching.	Near Threatened
Dolphins	Order: cetacean	Shallow areas of the oceans.	4–8 ft long and weighs 90–19.000 lbs; lifespan is 40–80 years.	Not available	Habitat loss	Endangered (ESA) listing

Name	Zoological Name	Habitat	Description	Present population	Threat	IUCN listing
Giraffes	Giraffa camelo-pardalis	Central, eastern and southern Africa	16 -19 feet tall. Weighing 1.600–3.000 pounds. Lifespan about 25 years.	Not available	Habitat destruction, pouching and fragmentation.	The West African or Nigerian Giraffe (G. c. peralta) and the Ugandan Giraffe (G. c. Rothschild) are endangered.
Frogs	Order: Anura 4740 species worldwide.	Highest concentration in warmer tropical climates.	Highly variable in size and weight.	Not available	Pollution, infectious diseases, habitat loss, invasive species, climate change and over – harvesting.	About one-third (32%) of the world's amphibian species are threatened or extinct.
Gorillas	Family: Hominidae, sub-species are Gorilla beringei grauer, Gorilla beringei beringe, Gorilla gorilla, Gorilla diehl.	Lowlands and mountains in SSA. Depending on the sub-species.	4,5–5,5 feet when upright, weighs 200–450 lbs, with a life-span of 30-50 years.	Not available	Pouching, habitat loss; Ebola disease, climate change.	Gorilla and gorilla beringei are critically endangered.

Name	Zoological Name	Habitat	Description	Present population	Threat	IUCN listing
Zebra	Equus qugga, Equus grevyi, Equus zebra	Savannas and mountains of many SSA countries depending on species.	3.5–5 feet at shoulders. 7–9 feet long. Weighing 440–990 lbs and with lifespan of 20–25 years.	About 4.500	Pouching, habitat loss, disease, climate change.	Grevy's zebra is endangered, mountain zebra is vulnerable.
Ring-tailed Lemur	Lemur catta	Highland country and thinly wooded forests of southern Madagascar.	Body is 15 to 18 inches long. Tails are 22–24 inches long. Weight: 5–8 pounds. Lifespan: 20 years.		Hunting, habitat loss to fires, overgrazing of domestic live-stock and logging.	Near Threatened
Orca or "killer whale"	Orcinus orca	Coastal waters and open ocean.	21–23 feet long. Weight: 4–10 lbs. Lifespan: 30–50 years.	50.000 world-wide	Pollution and chemical contamination.	Endangered (ESA listing)
African lion	Panthera leo	South Sahara desert and in parts of southern and eastern Africa.	Height: 4 feet. Length: 5–8 feet. Weight 330–500 lbs. Lifespan: 10–14 years. Top speed: 50 mph.	Fewer than 21.000	Hunting, pouching, habitat loss, disease, distemper, climate change.	Vulnerable

Name	Zoological Name	Habitat	Description	Present population	Threat	IUCN listing
Leopard	Panthera pardus	Forests, mountains, grass-land and deserts.	1.5–2.6 feet tall at the shoulder. Length: 3 -6 feet long. Weight: 82–200 lbs. Lifespan up to 20 years.	Not available	Fur trade and habitat loss.	Near threatened
Peregrine falcon	Falco peregrines	Wide variety of habitats including urban cities, the tropics and desert.	Length: 15–21 lbs (wings-pan of 3.5 feet. Weight: About 2 lbs. Lifespan: 7–15/20 years	SSA population is not available.	DDT Dichlorodiphenyltri-chloroethane, contamination and predation.	Endangered (ESA listing).
Sharks	Superoder Selachinmorp ha	Aquatic habitats at various temperatures.	7- 8 inches to 50 feet in length. Life-span: 20–100 years.			
Rhinoce-roses	Diceror bicornis, Ceratotherium simum	Savannas to dense forests in tropical and subtropical regions.	Rhinos range in height from 4.5–6 feet tall. Weight: 750–8.000 lbs. Lifespan: 35 years.	9.900	Poaching	Critically endan-gered to near threatened.

Name	Zoological Name	Habitat	Description	Present population	Threat	IUCN listing
Snakes	Suborder Serpentes	Many habitats including in the water, forests and deserts.	Snake size varies to extremes by species, some species will live as long as 50 years.	Difficult to estimate.	Roads and habitat destruction.	Many species are listed from critically endangered to extinct.
Whales	Order cetacean	Oceans	Length: up to 110 feet. Weight: up to 150 tons. Lifespan: 20–80 years.	Varies with each species.	Climate change	Endangered or threatened (ESA listing).
Wolf			26–32 inches at the shoulder. 4.5–6.5 feet from nose to tip of tail. Weight: 55–130 lbs. Lifespan: 7–more than 10 years.	200.000 world-wide.	Conflict with people over livestock losses.	Not available

Impact of climate change on endangered animal species in SSA

There are many factors that can cause a species to become threatened or endangered. The factors include habitat loss, pollution, competition from other species, disease, and unregulated killing and climate change. The contribution of climate change to species' loss is generally underestimated. According to Kotiaho *et al.* (2005), climate change has led to the endangerment of several species. A review across 14 biodiversity research centers in 2003 predicted that 15–37% of land species would be "committed to extinction" by 2050 due to climate change (Bhattacharya, 2004; Thomas *et al.*, 2004).

The process of global warming is already being witnessed in many parts of SSA and it includes vulnerability of the forest to fire, disappearance of glacier on African mountains such as Kilimanjaro, Ruwenzori and Kenya. Climate change also results in animal species range shift, change in forest ecosystems as well as tree and forage productivity. Changes in temperature, moisture levels, carbon dioxide and ultraviolet radiation may be more detrimental to forages, shrubs and water balance and can be implicated as the cause of food shortages, weight loss, endangerment and extinction of animals in SSA. Climate change now represents a great threat to the number of species surviving on earth. A large number of animal species in SSA have already been destroyed due to habitat-destruction and modification. No ecosystem on earth is immune to the effects of climate change. The impacts of global warming pose two problems across all habitat types, these are changes in phenology (timing of season elements) and changes in distribution of wildlife (Kostyack and Rohlf, 2008). Kotiaho *et al.* (2005) reported that climate change has led to the endangerment of several species. According to Roach (2004), by 2050 rising temperatures could lead to the extinction of more than a million species. Drought is one of the deadliest effects of climate change. Africa will receive the worst of these effects. The low availability of forage resulting from drought forces the animals to congregate at the little remaining water sources thereby increasing disease outbreaks of animals like the Grevyzebra foal.

Although change is a natural part of our evolving world, changes that occur at a pace that is too rapid usually have a disastrous effect on individual species of plants and animals. Habitat loss at an unprecedented pace is the primary cause of the endangerment of species in Africa. Climate and other factors related to habitat loss are posing serious threats to African wild dogs, cheetah, chimpanzee, giraffes, and gorillas. For bats that depend on nectar or fruits, changes in plant/flowering timing could put them out of synchrony.

Climate change projections also suggested that key portions of elephants' habitat will become significantly hotter and drier, resulting in poorer foraging conditions and threatening calf survival. Sharks, dolphins and whales are increasingly threatened by climate changes that are altering ocean circulation, sea surface temperatures, and even the chemistry and salinity of the ocean.

Protection of endangered animal species in Sub-Saharan Africa

Diversity of life and living systems are a necessary condition for human development (Ishwaran and Erdelen, 2006). Preservation of endangered animal species of SSA is important, not only because many of these species do and will continue to provide economic benefits in many part of SSA with large game reserves, but also because they constitute an important part of our ecosystem and offer a vast genetic "library" from which we can withdraw many useful items in the future. According to Bry (1986), habitat protection is the key to protecting our rare, threatened and endangered species. The first priority should be a comprehensive compilation of native animal species in SSA countries as well as the status of each of them with respect to whether they are endangered or not.

There is the need for the creation of a professional society dedicated to ecological ethics to help in research and management of biodiversity (Minteer and Collins, 2005) and to reduce the paucity of information on quantitative stimulation studies for estimating future climatic change impacts on endangered animals.

The above steps should be followed by management efforts that ensure that the habitat is intact in the face of climate change. The report of Douglas *et al.* (2004) contains recommended management actions such as: protection of coastal wetlands, accommodation of sea level rise, acquisition of inland buffer zones to provide an opportunity for wildlife to move inland, adjustment of yield and harvest models, building of corridors for wildlife movement, consideration of global warming models as well as historical data when making projections, looking for new habitats for wildlife, health maintenance health, genetically diverse populations, reduction of non-climate stressors on ecosystems, prevention and control of invasive species, reduction of the risk of catastrophic fires and the use of monitoring and adaptive management. Ishwaran and Erdelen (2006) submitted that the creation of biological corridors, biosphere reserves, ecosystem management, national monuments and eco-regional planning as approaches to integrate biodiversity conservation and socioeconomic development at increasingly larger spatial scales is a necessary step. Captive breeding – the process of breeding rare or endangered species in human controlled environments with restricted settings, such as wildlife reserves and other conservation facilities is recommended. The process saves species from going extinct, even though the techniques are reportedly difficult to implement for highly mobile species like some migratory birds and fishes. The need for new legal mechanisms in form of Endangered Species Act cannot be over emphasized in curtailing human-induced climate change problems. The fate of wildlife and ecosystems hinges on the willingness and ability of policy-makers and natural resource managers to confront global warming-related threats in Sub-Saharan Africa.

References

AFP, 2005. UN food agency warns Sub-Saharan Africa risks most from global warming. Rome (AFP), May 26, 2005.

Anonymous, 2009. Captive breeding populations – National Zoo FONZ. Nationalzoo.si.edu. http://nationalzoo.si.edu/ConservationAndScience/.

Bhattacharya, S., 2004. "Global warming threatens millions of species". New Scientist. http://www.newscientist.com/article/dn4545-global-warming-threatens-millions-of-species.html. Retrieved 2010-05-28.

Bramov, A., Belant, J. and C. Wozencraft, 2009. Gulo gulo. In: IUCN 2009. IUCN Red List of Threatened Species. Version 2009.2. www.iucnredlist.org Retrieved on 2010-01-25.

Bry, Ed., 1986. The rare ones. North Dakota Outdoors 49(2):2–33. Jamestown, ND: Northern Prairie Wildlife Research Center Home Page. http://www.npwrc.usgs.gov/resource/othrdata/rareone/rareone.htm Version 16 JUL 97).

Inkley, Douglas B. *et al.*, 2004. Global climate change and wildlife in North America (Wildlife Society, 2004).

International Union for Conservation of Nature, 2006. IUCN Red-list statistics. www.iucnredlist.org/info/tables.

IPCC, Climate Change, 2007. Impacts, adaptation and vulnerability: Contribution of Working Group II to the Fourth Assessment Report of the Intergovernmental Panel on Climate Change (Martin L. Parry *et al.*, eds., Cambridge Univ. Press).

Ishwaran, N. and W. Erdelen, 2005. Biodiversity futures, frontiers in ecology and the environment, 3(4), 179. Retrieved September 23, 2008. Ishwaran and Erdelen, 2006.

IUCN Red-list statistics, 2006. www.iucnredlist.org/info/tables.

Kostyack, J. and D. Rohlf, 2008. Conserving endangered species in an era of global warming. 4-2008 38 ELR 10203. Copyright © 2008, Environmental Law Institute®, Washington.

Kotiaho, J. S., Kaitala, V., Komonen, A., Päivinen, J. P. and P. R. Ehrlich, 2005. Predicting the risk of extinction from shared ecological characteristics, proceedings of the National Academy of Sciences of the United States of America, 102(6), 1963–1967. Retrieved September 24, 2008.

Minteer, B. A. and J. P. Collins, 2005. Why we need an "Ecological Ethics". Frontiers in Ecology and the Environment, 3(6), 332–337. Retrieved September 22, 2008.

Roach, J., 2004. By 2050 warming to doom million species, study says. National Geographic. News. http://news.nationalgeographic.com/news/2004/01/0107040107_extinction html.

The Independent, 2008. He's black, and he's back! Private enterprise saves southern Africa's rhino from extinction, The Independent, June 17, 2008.

Thomas, C. D. *et al.*, 2004. "Extinction risk from climate change". Nature (journal) 427 (6970): 145–148. doi:10.1038/nature02121. PMID 14712274. http://www.nature.com/nature/journal/v427/n6970/abs/nature02121.html. Retrieved 2010-05-28.

Ulansey, D., 2007. "The current mass extinction" repeats this statement with links to dozens of news reports on the phenomenon. Retrieved 2007.26.02. Endangered Species/CapBreedPops/default.cfm. Retrieved 2009-12-06.

UNEP/GRID, 2008. Vital Climate Graphics Africa, http://www.grida.no/general/2832.aspx.

Ndenecho Emmanuel Neba
University of Yaounde, Cameroon and
Lambi Cornelius Mbifung
University of Buea, Cameroon

Climate change and montane biodiversity management in Sub-Saharan Africa: A Study of the Cameroon highlands ecoregion

Abstract
The montane forests of Cameroon like most afro-montane forest refugia in Africa are unique. They have one of the highest levels of endemism particularly among birds and vascular plants. The biodiversity is extraordinarily rich; but despite this richness and their scientific importance, they have received little conservation attention and so have suffered from degradation. This study examines the biodiversity features, makes observations on current status of the forests, classifies the threats and establishes how climate change affects biodiversity. It concludes that climate change has major adverse impact on biodiversity, and consequently on ecosystem services such as food and water availability. The conservation and diversification of genetic resources are shown to play a fundamental role in mitigating climate change, buffering the influence of climate change as well as supporting rural communities of Sub-Saharan Africa (SSA) in adapting to it. This study sets the scope for adapting biodiversity management, food security livelihood sustenance and poverty alleviation strategies to climate change.

Introduction

The Cameroon highlands possess unique floral and faunal populations, with numerous endemics particularly among the birds and vascular plants. The high and often much localized degree of endemism is believed to have arisen during the Pleistocene age when climatic changes forced the forest to retreat to the wetter mountainous areas of the African continent (Figure 1), leaving tracts of forest or "refugia" cut off from one another. Within Cameroon, the most important refugia are thought to have existed in the areas surrounding Mount Cameroon, Mount Kupe and the Bamenda Highlands (Figure 2). The history of the distribution and evolution of montane forest species has been reviewed by Moreau (1966), Hamilton (1976, 1982), Diamond and Hamilton (1980), and Stuart (1986). It is now generally believed that past climatic changes have been instrumental in determining the present distribution of forest species, particularly of birds (Moreau 1966; Hamilton 1976). Prior to 12,000 B. P. the climate seemed to have been cold and dry. This must have resulted in the distribution of montane forests in the Cameroon Highlands, which were much different from that seen today. Since it was much colder, it is likely that the montane forests were displaced to lower altitudes. Since it was also drier, these forests were probably more restricted than they are today (Livingston 1975; Hamilton 1982). It is generally believed that the wettest places today are likely to be the wettest places during that period (Livingston, 1975). In West Cameroon, the wettest places are Mount Cameroon, Mount Kupe, Mount Nlonako and the Bamenda Highlands (Figure 2). It is therefore likely that montane forests were restricted to these lower mountain slopes during the last glaciations and became important forest refugia for many montane species of plants and animals.

The restricted and isolated distribution of small populations of montane flora and fauna in forest refugia is believed to have resulted in specification and subspecification, since the period of isolation was fairly brief (around 20,000 years before the temperatures rose and the forest expanded), most differentiation was at subspecies level. Many of the montane forest species have been unable to disperse subsequently from the refugia to colonize other mountains.

Figure 1: The distribution of afromontane forest refugia in Africa.

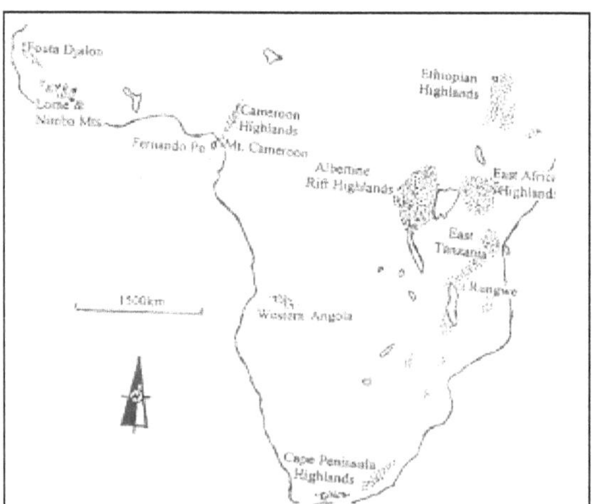

When the climate became much milder and humid, about 12,000 years ago, montane forests were probably displaced up to higher elevations by lowland forest, while increasing humidity made it possible for the forest to expand beyond the refugia to mountains previously too dry to support forest growth. The expansion of the forest could have increased the opportunities of colonization for long distance wanderers. This could explain the disjointed distribution of afro-montane vegetation and birds in East and West Africa. Whatever the explanation, the Cameroon Highlands exhibit the western limits of typical members of East African fauna and possess a vegetation with many East African affinities, particularly the *Podocarpus/Arundinaria* formation which is widespread in East Africa, but now found nowhere else in West Africa (Macleod, 1986). The paper examines the biodiversity features, makes observations on the current status of the forest, the types and severity of threats and establishes how climate change affects biodiversity. It finally, identifies the scope for mitigating climate change as well as supporting traditional societies in Sub-Saharan Africa (SSA) in adapting to it.

Figure 2: Location of the West Cameroon Highlands.

Geographical setting of the eco-region

Geographically, this ecoregion encompasses the mountains and highland areas of the border region between Nigeria and Cameroon, excluding Mount Cameroon. Most of the ecoregion lies within a narrow rectangle of some 180 km by 625 km, oriented southwest to northeast and originating about 50 km inland of Mount Cameroon. At the southern extremity, the ecoregion covers the Rumpi Hills, the Bakossi Mountains, Mount Nlonako, Mount Kupe and Mount Manengouba. It then extends northeast towards the Mambila Plateau, with an extension north to the Bamenda-Banso highlands and outliers northeast to Mambila and northwest to the Obudu Plateau of Nigeria. The main trend of the ecoregion continues north-east along the western flank of the Adamawa Plateau to Tchabal Gangdaba with two small outliers further east (Stuart, 1986; Gartlan, 1989; Stattersfleld *et al.*, 1998). The highest point within this eco-region is Mount Oku at 3,011 m in the Bamenda-Banso highlands. Most of the remainder is below 2,600 m in elevation. At about 800 m to 1,000 m the ecoregion grades into lowland vegetation communities of other ecoregions. In the majority of cases, however, the lower boundary of these forests is now determined by conversions to agriculture and other land-uses. Although located in tropical Africa the mean maximum temperatures are below 20°C due to the effects of altitude. At the southern extremity, closer to the coast, rainfall is around 4,000 mm per annum, declining inland to 1,800 mm or less.

These mountains form part of a chain of former volcanoes that stretch inland from the sea. While there are currently no active volcanoes within the eco-region, the effects of volcanism are still present. Soils derived from these volcanoes are fertile, which makes the land attractive to farmers. Combined with adequate rainfall, this contributes to a high human population density. In White's (1983) phytogeographical classification, these mountain areas fall within the Afromontane archipelago – like regional center of endemism that spans the entire continent. The elevation stratification that occurs along these mountains is quite distinct. Vegetation consists of sub-montane forests between 900 m and 1,800 m and above this is a mixture of montane elements, including distinct montane forests and patches of montane grasslands, bamboo forests and subalpine communities. Five species of trees characterize the forested montane zone: *Nuxia congesta, Podocarpus latifolius, Prunus africana, Raparzea melanophloeos,* and *Syzygium guineense bamendae,* and these trees become increasingly covered with an epiphytic flora, especially orchids and mosses at higher altitudes. Other important montane species include *Crassocephaluim mamnii, Hypericum lanceolatum, Myrica humilis, Piilippia mannii, and Schefflera abyssinica.*

Observations on biodiversity features

Although plant diversity and biogeography are poorly known, tree species diversity tends to be low, but the diversity of non-woody plants such as grasses is high. Highest levels of tree endemism are found in the submontane region and the higher elevations of the adjacent Mount Cameroon (Figure 2) ecoregion. Overall, tree diversity, however, decreases with increasing elevation. There is also a significant endemic flora in the grasslands, heath lands, moorlands and other non-forested habitats at higher elevations. These non-forested habitats share elements with the high mountain plant communities to the east, namely the Ruwenzori-Virunga montane moorlands. The Cameroon Highlands ecoregion is not high enough to contain pure Afroalpine elements, which are restricted to the highlands of East Africa (Figure l).

The forests have affinities with the highland forests of Angola on the southern fringes of the Congo Basin (Kingdon, 1990) and especially with forests in East Africa. Most of the trees in the montane forests of Cameroon are also present in the mountains of eastern Africa; some examples include *Alangium chinense, Albizia gummifera, and Apodytes dimidiata. Cassipourea gummiflua, Croton macrostachyus, Ilex mitis, Olea capensis, Podocarpus latifolius, Polysdas fulva, Prunus africana, Schefflera abyssinica, Strombosia scheffleri, Xymalos monospora,* and at edges *Agauria salicfolia, Maesa lanceolata, Myrica humilis, Nuxia congesta, Pittosporum Sp., Rapanea melanophloeos and Scolopia zeyheri* (Dowsett-Lemaire, 1989). The liane *Dregea schimperi* (Asclepiadaceae), common in the drier type of montane forests of this ecoregion, is widespread in the mountains of East Africa (Thomas and Thomas, 1996).

The eco-region contains exceptional levels of avian endemism. There are seven strictly endemic bird species, the Bamenda apalis *(Apalis bamendae),* Bangwa forest warbler *(Bradypterus bangwaensis),* white-throated mountain-babbler *(Kupeornis gilberti,* EN), banded wattle-eye *(Platysteira laticincta,* EN), Bannerman's weaver *(Ploceus bannermani,* VU), Bannerman's turaco *(Tauraco bannermani,* EN), and Mt. Kupe bushshrike *(Telophorus kupeensis,* EN) (Bowden and Andrews, 1994; Stattersfield *et al.*, 1998). An additional nine montane endemics are in common with Mt Cameroon and Bioko: These are *Psalidoprocne fuliginosa, Andropadus tephrolaemus, Phyllastrephus poensis, Phylloscopus herberti, Urolais epichlora, Poliolais lopezi, Nectarinia oritis, Nectarinia ursulae, Nesocharis shelleyi.* There are also 14 species in common with just Mt Cameroon and not Bioko, adding *Andropadus montanus, Phyllastrephus poliocephalus, Laniarius atroflavus, Malaconotus gladiator, Cossypha isabellae* and the race

Cisticola chubbi discolor (sometimes considered a separate species C. *discolor)* (Dowsett, 1989; Dowsett-Lemaire and Dowsett, 1989, 2000; Fotso *et al.*, 2001).

Eleven small mammal species are considered strictly endemic to this region: Eisentraut's striped mouse *(Hybomys eisentrauti,* EN), an African wood mouse species *(Hylomyscus grandis),* Mount Oku mouse *(Lamottemys okuensis,* EN), Mittendorf's striped grass mouse *(Lemniscomys mittendorfi,* EN), *Lophuromys dieterleni* and *L. eisentrauti,* Oku mouse shrew *(Myosorex okuensis,* VU), Rumpi mouse shrew *(M. rumpii,* CR), western vlei rat *(Otomys occidentalis,* EN), Hartwig's soft-furred mouse *(Praomys hartwigi,* EN), and Isabella's shrew *(Sylvisorex isabellae,* VU). In addition to these smaller species, there is also an isolated population of an endemic subspecies of lowland gorilla *(Gorilla gorilla diehli,* EN). There are also populations of rare larger mammals – for example there are several groups of drill *(Mandrillus leucophaeus,* EN) in Bakossi, as well as healthy populations of Preuss's red colobus *(Procolobus pennanti preussi,* EN) and chimpanzee *(Pan troglodytes,* EN). One large group of drill was seen on Mount Kupe as recently as the year 2000.

Very high levels of endemism are observed among amphibians, with nearly 40 species as strict endemics (Gartshore in Stuart, 1986): *Hylarana longipes, Petropedetes parkeri, P. perreti, Phrynobatrachius cricogaster, P. steindachneri, P. werneri, Phrynobatrachus sp.* (Oku), *Phrynodon Sp.* 1 *sensu* Amiet (1975), *Phrynodon sp.* 2 *sensu* Amiet, *Arthroieptis adoififriedericii, Cardioglossa melanogaster, C. oreas.*

Among the reptiles the following nine species are considered narrow endemics: Atractapsis coalescens, Chamaeleo eisentrauti, Pfeffer's chameleon (C. pfefferi), four-horned chameleon (C. quadricorrnis), C. nemaspis gigas, Leptosiaphos chrisimldi, L. ianthinoxantha, angel's five-toed skink (L. lepesmei), and Panaspis duruarum. In addition to the narrow endemics, there is also a significant overlap between the flora and fauna of this eco-region and that of the nearby Mount Cameroon and Bioko Montane Forests eco-region. Fifty near endemic plant species are shared between the two eco-regions. Similarly, 27 bird species are confined to these two eco-regions (Stattersfield *et al.*, 1998). A similar pattern of overlap in taxa is seen in the plants, birds, reptiles, amphibians, invertebrates, and small mammals.

The current montane biodiversity status, types and threats

While the different mountains and highlands that comprise this ecoregion are naturally geographically isolated, human activities are increasingly fragmenting, degrading and isolating the remaining forest patches. The area around Mount Okuin the Bamenda-Banso Highlands supports some 100 km^2, and there are other patches in this region. The Bakossi Mountains have at least 200 km^2 of mid-altitude and montane forest above the altitude of 1000 m; and the lowland forest ("Western Bakossi") covers some 400 km^2. The Mt. Nlonako Faunal Reserve also contains a partial forest continuum from the montane section to the lower levels, Tchabal Mbabo, in the northern sector, has almost 50 km^2 of virtually pristine montane forest (Thomas and Thomas, 1996). In Nigeria, the biggest patch in the Gotel Mountains covers 46 km^2 (peak Gangirwal) and there are other patches. Montane forest remnants also remain in gullies of the Obudu Plateau and on the highland areas further to the north such as the Mambila and Mana Plateaux and Gashaka Gumti; all are smaller than the Cameroon patches (Sayer *et al.*, 1992; Stattersfield *et al.*, 1998).

Forest loss has been significant on many mountains in this eco-region. Much of the Bamenda-Banso Highlands were once covered by forest, but the cover has declined by more than 50% since the 1960s through conversion for agricultural use, because of the relatively fertile soils and the reliable rainfall in this area (Stuart, 1986; Collar and Stuart, 1988; Alpert, 1993). Similarly, deforestation has also occurred on the Obudu Plateau. This is one of the least well-protected ecoregions in Africa. No part of this ecoregion is under formal protection status in Cameroon, although local traditional rulers still exert considerable authority over land use. The main section of Bakossi (550 km^2) has been proposed as "Protection Forest", banning all logging. Kupe has been proposed as a "Strict Nature Reserve", and the boundaries of this reserve were successfully delineated with the participation of the local people in 2000–2001. The Kilum forest at Oku has some form of protection and the boundaries are well demarcated; unfortunately, the demand for firewood for neighboring towns (including Bamenda) is so high that the forest may not have a sustained future. Nlonako Mountain is a Faunal Reserve, which gives it protection from logging, even if the hunting ban is not respected. In Nigeria, the Gashaka Gumti National Park contains some montane forest and some fragments remain at Obudu in the Okwango section of the Cross River National Park.

The natural habitats of this ecoregion are highly threatened and are being lost through conversion to agriculture, unsustainable use of timber, fires from farmland, and the collection of firewood and construction materials (Collar and Stuart, 1988; Gartlan, 1989; Alpert, 1993). Because volcanic rock produces good soils, there is considerable pressure to convert such volcanic areas to farmlands. However, these soils also dry easily, leading to desiccation during the dry season. Unsustainable collection of firewood is a major cause of degradation of forest patches. Hunting also threatens the remaining larger mammals.

Several plants and animals are used for food in traditional communities. Diversity is important for nutritionally adequate diets. Adaptation to changing conditions due to climate change is better achieved through diversification, reducing the negative impacts of extreme events, like floods, droughts, or diseases. In general, wild sources of food remain particularly important for the poor and landless and play a very significant role in times of famine, and insecurity and conflict, when food supply mechanisms are disrupted and people have limited access to food. Traditional communities are also highly dependent on ethnopharmacology. Several medicinal plants have been identified by the Limbe Botanic Garden. Biodiversity contributes not only directly to food security, but also to a number of environmental services provided to agriculture, such as pollination of food crops, water regulation, disease control and others. Intact ecosystems are important sources of local livelihood. Biodiversity also serves as a safety net for vulnerable households in time of crisis, because it offers income opportunities, and sustains productive agricultural ecosystems. Development actions must focus on the functional relationships and processes within ecosystems. A key question is "how local livelihood improvement and food security can be linked to biodiversity conservation and land management in the buffer zones of protected areas?"

Climate change related impacts on the biodiversity

Climate change has a wide range of effects on biodiversity. Plants and animals rely on specific environmental conditions; changes in temperature and humidity regimes affect the performance and competitiveness and relationships of species. Species adapted to cold conditions, retreat to higher altitudes and latitudes while species living in warm areas may expand their ranges of distribution. This can cause problems if species expand to levels that are regarded as pests, or if invasive species do harm to locally adapted ecosystems. The desert locust is an example, a slight increase in precipitation in susceptible areas in Africa where this insect lives may cause increased population densities (Cheke and Tratalos, 2007, cited by Vohland and Lotze-Campen, 2008).

However, shifts in distribution areas will not always be possible as species may not be able to migrate through a landscape which has been increasingly fragmented by human activities (Figures 3, 4, and 5). They may not also find a new place to establish as climate regimes may not support it or land use is inadequate. Consequently climate change would be expected to further increase the global rate of species extinction.

The rapid change of environmental conditions is of particular concern for some hotspots of species diversity in Africa. These areas could transform from tropical forests to a Savannah-like formation. These areas will then lose large numbers of species, many of which are still unknown. These species form part of the living library needed to detect new pharmacological or crop species.

Figure 3: Degradation of Mount Oku Forest. Source: Macleod, 1986.

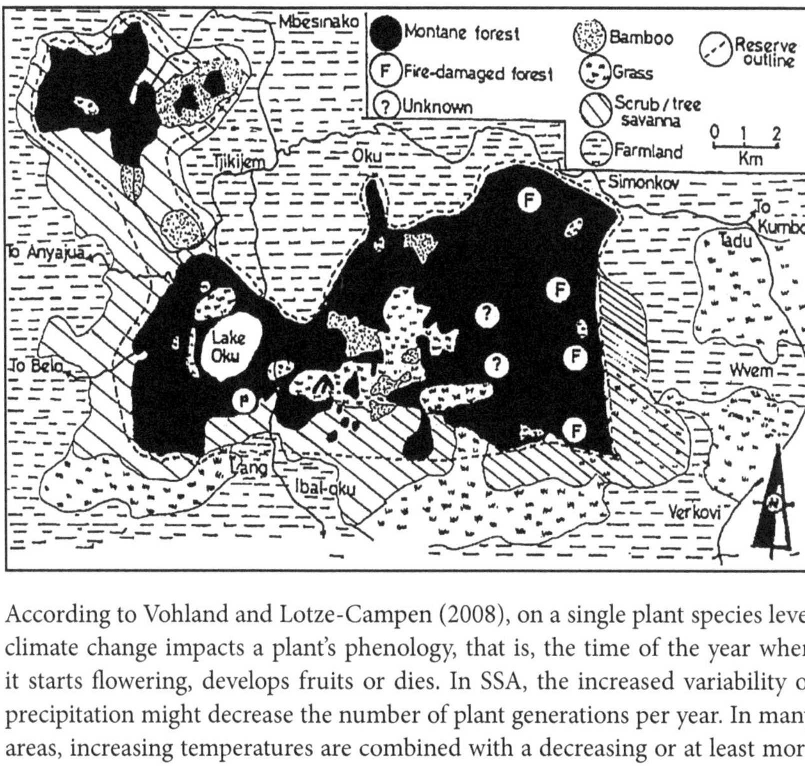

According to Vohland and Lotze-Campen (2008), on a single plant species level climate change impacts a plant's phenology, that is, the time of the year when it starts flowering, develops fruits or dies. In SSA, the increased variability of precipitation might decrease the number of plant generations per year. In many areas, increasing temperatures are combined with a decreasing or at least more variable supply of water.

Another climate change related impact described by Vohland and Lothe-Canpen (2008) is the potential "fertilization effect" of atmospheric carbon-dioxide (CO_2) increases. Many plants are limited in their growth by CO_2; therefore increased supply may lead to increased biomass accumulation or enhanced phenological developments as long as water and other nutrients are not limiting. In principle this is true for crops, too, but recent research results indicate that the delivery value of some crops might be reduced as the supplementary CO_2 leads to more sugar instead of proteins (Stafford, 2007, cited by Vohland and Lothe-Campen, 2008).

Decreasing biodiversity may be related to reduced food security as varying conditions cannot always be buffered by genetic or species diversity. The risk of famine increases with climate changes, particularly when faming systems and livelihood options rely on a few or only one species and institutions cannot provide support. Clearing mountain forest as shown in Figures 3, 4 and 5 reduce the water storage functionality in catchments. In combination with floods and droughts induced by climate change, people suffer from soil erosion and water depletion.

Figure 4: Degradation of Mount Kupe Forest.

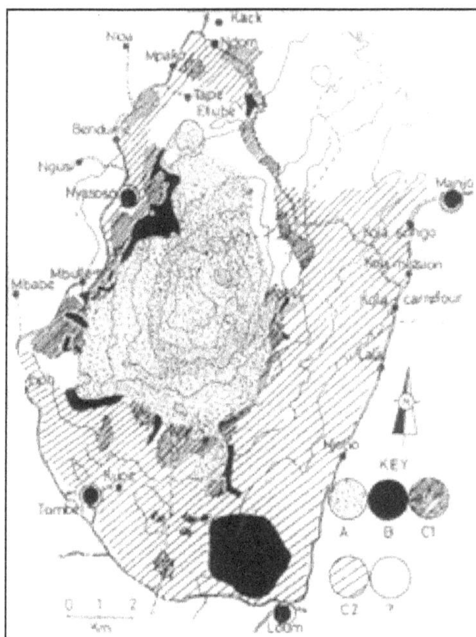

A = primary forest, B = secondary or disturbed forest, C1 = farms with less than 50% canopy cover, C2 = farms with a few or more trees, ? = areas yet to be mapped. Source: Mount Cameroon Project Report.

Implications for rural development

The problems presented by climate related impacts are challenges for rural development and poverty alleviation. The biggest challenge is to tackle increasing rainfall variability, that is, the problem of droughts and floods combined with the overall loss of biodiversity. These are not only caused by climate change. Key areas for adaptation to climate change include;

- The design of guidelines for bio-energy production: Rural areas in SSA face a new period of agricultural intensification for a rapidly growing population and the production of biomass for energy. In order to save forest refugia, bio-energy production might support rural development and maintain biodiversity on a small scale, especially on a farmer-scale.
- Protection of forest and trees. One of the key roles of vegetation with respect to climate change is to buffer against increasing temperatures.
- Respiration of forest enhances the cooling of the surrounding area and may even increase the probability of precipitation. The delineation of areas for protection forests should have the following considerations:
- Delineation of montane forest areas for protection above 900 m elevations for capturing sub-montane forest habitats and sub-alpine communities.

Figure 5: Conservation values and threats in the Mount Cameroon Region.

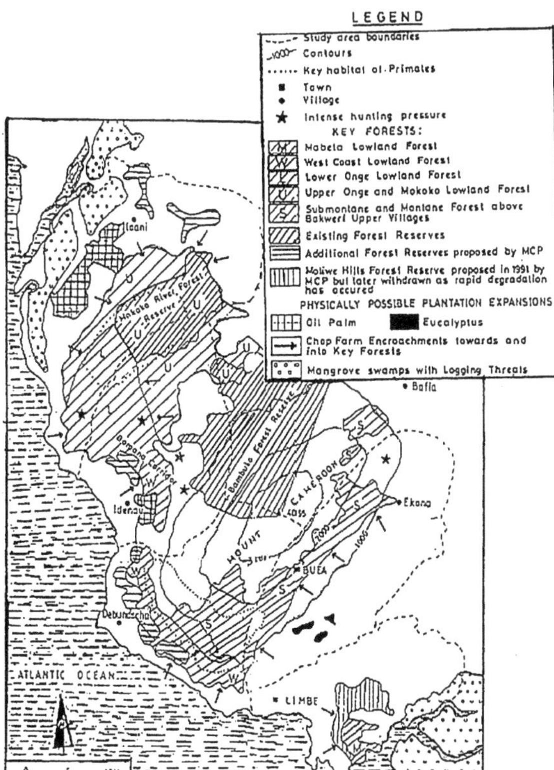

MCP = Mount Cameroon Project, Source: Mount Cameroon Project

- Delineation of protection forest in the lower elevations in order to capture elevation migration of species as well as transitional zones that appear to be crucial component of speciation and of the development of endemism (Schneider *et al.*, 1999).
- The development of agro-forestry systems: Agro-forestry systems combine biodiversity conservation and climate change adaptation. Agro-forestry systems are more resilient against climate change than treeless fields, as the trees reduce wind speed and therefore evaporation, and trap water and nutrients. They also harbor higher species diversity including predators that check insect pests.
- Enhancement of genetic crop diversity: This involves the development of a high agro-biodiversity which allows for local adaptation processes. It will require that farmers be supported in maintaining their locally adapted native varieties as an insurance against future increases of temperature variation and the occurrence of new pest and diseases.
- Development and promotion of soil and water conservation strategies: soil tillage in combination with high temperatures leads to a rapid oxidation and loss of soil carbon. Organic farming and soil conservation practices such as mulching or zero tillage enhance both the soil biological diversity, its fauna and also reduce carbon losses from the soil. Paying attention to the conservation of soils will permit a more sustainable agriculture to face climate change.
- Promotion of biodiversity governance at the global, regional and national levels. Biodiversity governance must be an intrinsic part of international and national legislation. Protected areas as a means to conserve the library for future crops and pharmacological development as well as retreat areas for plants and animals must be supported by international funding and trading mechanisms.

  Nature conservation funds and "payments for ecosystem services" in SSA should be a global task; it should not be regarded as the responsibilities of the countries concerned. Such payments should encourage private land users to introduce environmentally friendly land use practices. The aim should be to ensure that sustainable land resource management and conservation apply not only to specially designated protected areas, but also on productive land. The concept is based on the innovative idea of making payments for ecosystem services. This is particularly important, because SSA faces additional challenges such as poverty and livelihoods sustenance:
- Poverty exacerbates competition for scarce resources and means that natural resource conservation and biodiversity protection have relatively low priority.

- The creation of protected areas is usually associated with high opportunity costs, because these areas could often also be used for food production and other livelihood activities.

Conclusion

The Cameroon Highlands eco-region comprises montane forest/grassland patches mainly above 900m elevation, scattered in an archipelago along the border area between Nigeria and Cameroon. Habitat varies with increasing altitude from sub-montane to montane and ultimately sub-alpine grasslands. The forest and grasslands contain exceptional levels of endemism in all taxa. In most areas, these are poorly protected. Due to population pressure the remaining habitats, especially the forest, are threatened by agricultural expansion. Climate change also has a major impact on the biodiversity. Sustainable biodiversity management through the measures outlined above can play an important role in mitigating climatic change, buffering the influence of climate change as well as supporting rural communities in Sub-Saharan Africa in adapting to it.

Acknowledgements

The authors acknowledge the use of descriptions of the Cameroonians Highlands forests (ATO103) of Wild World: www.Worldwildlife.org.

References

Alpert, P., 1993. Conversing biodiversity in Cameroon. Ambio 22:44–49.

Amiet, J. L., 1975. Ecologie et distribution des Amphibiens Anoures de la region de Nkongsamba (Cameroun). Ann. Fac. Sei. Yaounde 20:33–107.

Amiet, J. L. and F. Dowsett-Lemaire, 2000. Un nouveau Leptodactylodon de la Dorsale Cameroonaise (Amphibia, Anura). Alytes 18:1–14.

Bowden, C. G. R. and S. M. Andrews, 1994. Mount Kupe and its birds. Bulletin African Bird Club 1:13:16.

Collar, N. J. and S. N. Stuart, 1988. Key forests for threatened birds in Africa. ICBP, Cambridge, UK.

Diamond, A. W. and A. C. Hamilton, 1980. The distribution of forest passerine birds and quaternary climate change in tropical Africa. Jour. Zool. London. Pp. 379–402.

Dowsett, R. J. (ed.), 1989. A preliminary natural history survey of Mambilla Plateau and some lowland forest of eastern Nigeria. Tauraco Research Report, No. 1, 56 pp. Tauraco Press.

Dowsett-Lemaire, F., 1989. The flora and phytogeography of the evergreen forests of Malawi. 1: Afromontane and mid-altitude forests. Bull. Jard. Bot. Nat. Belg. 59:3–131.

Dowsett-Lemaire, F. and R. J. Dowsette, 1998. Zoological surveys of small mammals, birds and frogs in the Bakossi and Kupe Mts. Cameroun. Unpublished report for WWF-Cameroon, 46 pp.

Dowsett-Lemaire, F. and R. J. Dowsette, 2000. Further biological surveys of Manenguba and Central Bakossi in March 2000 and an evaluation of the conservation importance of Manenguba, Bakossi, Kupe, and Nlonko Mts. with special reference to birds. Unpublished report for WWF-Cameroon, 45 pp.

Fotso, R., Dowsett-Lemaire, F. and R. J. Dowsette, Cameroun Ornithological Club (P. Schölte, Languy, M. and C. Bowden, 2001). Cameroon. Pp. 133–159, In: Fishpool, L. D. C. and M. I. Evans (eds.). Important birds' areas of Africa and associated islands: Priority sites for conservation. Bird Life Conservation Series No. 11. Pisces Publication and Bird Life International, Newburg and Cambridge, UK.

Gatland, S., 1989. La conservation des ecosystems forestiers du Cameroun. IUCN, Gland and Cambridge.

Hamilton, A. C., 1976. The significance of patterns of distribution shown by forest plants and animals in tropical Africa for the reconstruction of Upper Pleistocene Palaeo-environment: A review. Paleoecol. Afr. Vol. 9, 83–97.

Hamilton, A. C., 1982. Environmental history of East Africa: A study of the Quaternary. Academic Press, London.

Kingdon, J., 1990. Island Africa; the evolution of Africa's rare animals and plants. Collins, London.

Livingston, D. A., 1975. Late quaternary climate change in Africa. Ann. Rev. Ecol. Syst. Vol. 6, pp. 248–249.

Macleod, H., 1986. The conservation of Oku Mountain forest. ICBP, Cambridge, pp. 33 -35.

Moreau, R. E., 1966. The bird fauna of Africa and its islands. Academic Press, London. Mount Cameroun Project Report.

Sayer, J. A., Harcourt, C. S. and N. M. Collins, 1992. The Conservation Atlas of Tropical Forests: Africa. IUCN and Simon & Schuster.

Schneider, C. J., Smith, B., Larison, B. and C. Moritz, 1990. A test of alternative models of diversification in tropical rainforests: Ecological gradients vs. rainforest refugia. Proceedings of the National Academy of Sciences, Vol. 96, No. 24, pp. 13869–73.

Stattersfield, A. J., Crosby, Long and D. C. Wege, 1998. Endemic bird areas of the world: Priorities for biodiversity conservation. Bird Life Conservation Series, No. 7. Bird Life International, Cambridge, UK, International Council for Bird Preservation.

Stuart, S. N. (ed.), 1986. Conservation of Cameroon montane forests. Cambridge, UK. International Council for Bird Preservation.

Thomas, D. and J. Thomas, 1996. Tchabal Mbabo Botanical Survey. Consultants' Report to WWF Cameroon Program Office. 44 pp.

Vohland, K. and H. Lotze-Campen, 2008. Biodiversity and climate change. The International Journal for Rural Development. DLG-Verlag GmbH, Frankfurt.

White, F., 1983. The vegetation of Africa, a descriptive memoir to accompany the UNESCO/AETF AT/UNSO Vegetation Map of Africa (3 Plates, Northwestern Africa, Northeastern Africa and Southern Africa (1:5,000,000). UNESCO, Paris. Prepared by Allard Blom, Reviewed by: in progress.

Bernard P. K. Yerima
University of Dschang, Cameroon

Protective measures taken for preserving species endangered by climate change in Sub-Saharan Africa: Case of Yongka Western Highlands Research Garden-Park, Nkwen-Bamenda, Nw Region, Cameroon

Abstract

Africa's biodiversity and that of Cameroon in particular hold great potentials for transforming the continent's agriculture and directly contributing to economic growth and alleviating poverty situation in the region. This diversity is currently underutilized, neglected and continually eroded rather at alarming rates. In this paper status of work currently undertaken at the **Yongka Western Highlands Research Garden-Park, Nkwen-Bamenda, Cameroon is presented.** The park serves as a research and training centre for shaping innovation technologies that protect the watershed leading to increased and sustained water yields, providing timber for construction and wood for fuel, contributing to the reduction of the adverse effects of climate change and reducing pressures currently exerted on endangered plant species. The 4-ha Research Park is located in Nkwen- Bamenda, Cameroon at an altitude of about 1000 m above mean sea level with a marked wet and dry season. Soils are varied, acid in nature, have low nutrient contents and compact and dense structures that impede plant root growth. Research activities carried out reflect the multitude of specific objectives on themes which mirror the complexity and the interwoven and multi-disciplinary nature of the research problem. Evaluation of plant performance of selected plant species (fruit trees, medicinal plants, spices, etc.), is based on soil and water conservation and soil profile modification techniques that improve root penetration, nutrient content and water retention capacities. This work provides a framework for conserving biodiversity, most importantly plant species at risk of extinction, through the establishment of a Plant Gene Bank, and practical training and research as well as a medium for reproducibility and scaling up of these procedures to areas having similar challenges. Despite the promising nature of the results obtained so far, many challenges largely associated with financing research and outreach activities remain. Government's support and creation of an enabling environment to ensure conservation of genes from conservation of biodiversity through a land use plan regulatory framework and increased sensitization of the local community would sustain this effort and preserve endangered species. For sustainable use and conservation of biodiversity, countries of the Sub-Saharan Africa region will need to harness and apply science and technology, invest in research and innovation to generate technologies for conservation and sustainable use of the biodiversity.

Introduction

The biodiversity of the Sub-Saharan region of Africa holds great potentials for transforming the continent's agriculture system and contributing to economic growth, poverty reduction and improved standard of living of the people. The unique species of plants and animals as well as ecosystems constitute the continent's natural wealth. However, this plant diversity is underutilized and is gradually being eroded at alarming rates. Though conserving and promoting sustainable use of biodiversity is one of the challenges that African countries have committed themselves to addressing, as manifested by many countries that ratified the UN Convention on Biological Diversity, this commitment has yet to be translated into practical terms.

For sustainable use and conservation of biodiversity, African countries south of the Sahara will need to harness and apply science and technology as these activities are knowledge-intensive. This calls for investments in research and innovation to generate technologies for conservation and sustainable use of biodiversity. Plant Gene Banks aim to safely conserve plant diversity. Gene Banks cover many possibilities, from massive collections stored in elaborate buildings to simple fields of a few labeled plants. Gene Banks are *ex-situ* collections that are comprised of samples stored off site, away from the environments in which they naturally grow. Plant Gene Banks are like an insurance policy for the future of agriculture, medicinal plants and other endangered plant species. They play a crucial role in the conservation and use of biodiversity.

Plant genetic resources are the basis for the survival and well-being of human beings. We require them to increase food supplies and agricultural production sustainably and to meet the related challenges in the environment. Medicinal plants and traditional medicine play an important role in the health care system of most developing countries. Reports from Ethiopia indicate that about 80% of the human population and 90% of livestock rely on traditional medicine (IBC, 2009). This situation is very similar to that of Cameroon and other developing countries where over 60% of the population still relies on traditional medicine.

The importance of this resource notwithstanding, the medicinal plant biodiversity is being depleted due to man-made and natural calamities. In view of this preoccupying situation, some countries south of the Sahara are creating Field Gene Banks (FGB) both for *in-situ* and *ex-situ* conservation of endangered plant species at risk of disappearance. Notably among these are Ethiopia, South Africa, Burkina Faso and Ghana.

The situation in North West Cameroon

Over the years, due to increased demographic pressures, gallery forests in the North West region of Cameroon have been destroyed resulting in the total loss of prior existing plant species. In some areas the government has responded to these losses by creating natural forest reserves. Unfortunately, in the absence of appropriate control measures these reserves have been subject to massive poaching from an uncaring population. An example is the Bamenda Highlands area, which drains into the Bamenda urban area and presently has two forest reserves. The Bali Ngemba Forest Reserve positioned facing west in a concave part of the steep scarp slope that connects the High Lava Plateau (1500–2100 m alt.) of the Bamenda Highlands with the Low Lava Surface (900–1200 m alt.) on which the former appears to rest (Hawkins and Brunt, 1965) is made up of basalts and trachytes of ancient volcanic origin (Conrade, 1974). The Bafut Ngemba Forest Reserve lies to the east of the Bali Ngemba Forest Reserve, has similar geology and geomorphologic elements to the latter. It has a larger surface area and occupies a higher altitude. According to Hawkins and Brunts (1965) and Harvey *et al.* (2004), the vegetation in this area has undergone much degradation. Cheek *et al.* (2000) and Harvey *et al.* (2004) estimated that 96.5% of the original forest cover of the Bamenda Highlands above 1500 m altitude has been lost.

Thus, these two government forest reserves presently exist mostly in name, as most of the forest under storey has been cleared, and the fertile forest soils used for growing food crops and grazing. Government forest guards in this area are powerless to stop the carnage. The conservation of this site may be facilitated by its forest reserve status, but its future appears very bleak at the moment, except for forests on the most inaccessible slopes. Pressure on the forest reserves come from the population centers adjoining the forest reserves in the Bamendankwe, Bambili, Santa, Pinyin, Bafut, and Nkwen villages where small scale logging has systematically degraded the forest giving way to agriculture and pastoral activities. As the forest reserves are official government and not tribal (traditional) property, protection from the traditional authorities does not obtain (Harvey *et al.*, 2004). Instead, both forest reserves are tribal no-man's land, weakly guarded and so open to abuse. '*If we don't exploit the forest others will anyway*', seems to be the feeling on the ground, which is likely to persist until government can put in place real dissuasive measures that would make such contravention worthless (Harvey *et al.*, 2004).

Similar scenarios can be found all over the country. For example, in Dschang in the lake area adjoining the University of Dschang, where a High School has been constructed, a sign board drawing attention to the non-intrusion by the population on the government forest reserve, which existed in the past stand telling a sad story of the neglect/failure of government policy and the will to address the situation. The consequences of this neglect are immense and will come to haunt our future generations. Already in this area, the nearby lake and stream are drying up due to siltation from sediment resulting from sand quarrying, farming and construction activities in the adjoining uplands.

Historical perspective of biodiversity conservation in the Bamenda Highlands region

In pre-colonial time villages here lived as collective entities under powerful traditional chiefdoms with strong family ties. Successions of offspring and increased demographic pressures led to partition of land, resulting in fragmentation of the resource making sustainable management difficult. This has been hastened by urbanization with little accountability in contrast to the traditional set up that existed before. The key problems have been catalogued as soil erosion, siltation, flooding, depletion of soil fertility, destruction of watersheds, lowering of water tables and landslides, attributed to use of poor and in-adapted traditional shifting cultivation methods, excessive deforestation, cultivation on fragile (steep) hill-slopes, setting of bush fires, overgrazing by cattle, building of settlements, and increased consumption of fuel wood (UNDP, 1994; Yerima and Van Ranst, 2005 a, b). Fuel wood supplies over 90% of domestic energy in many African nations (Eckholm *et al.*, 1984). Adding a sense of urgency to the fuel wood problem is the estimate that 50 million Africans experienced acute fuel wood scarcity in 1980 and that supplies are still dwindling (FAO, 1985). This problem is very acute in the **Bamenda Highlands region**. Appropriate knowledge about the types of vegetation that would be most effective in bringing about the best ground coverage, leading to increased water yields and satisfying other economic needs such as fuel wood needs, timber for building, conservation of medicinal plants and other plants of economic value would be required.

Reforestation would protect the soil against erosion, regularize the water flow within the basin, reduce the effect of climate change and protect and improve the watershed through its influence on infiltration rate associated with increased interception and breakage of raindrops and reduced splash and runoff, aggregation of primary soil particles by organic matter, roots which act as conduits for water down the soil, and organisms which burrow the soil enabling water infiltration. Further, this activity would provide wood for fuel and timber for house construction, some of the problems confronting the local population in this area.

Many approaches are being envisaged by some countries south of the Sahara. For example, the Institute of Biodiversity Conservation of Ethiopia (IBC, 2009) observed that plant genetic resources for food and agricultural activities involve public and private institutions and companies, non-governmental organizations, communities and individuals from the conservation environment and development sectors. The latter institution also observed that the integration of existing activities within the framework of a unified national program at the national level provided the opportunity to enhance such diverse efforts in the country.

Objectives

The main research axes at Yonka Research Park include: biodiversity conservation, agro-forestry research; creation of a Plant Gene Bank; watershed management; adaptability of plant species to climate change; soil-water-plant related constraints; outreach of promising technologies through student study tours/research activities, farmer field days and visits by decision makers, that would ensure sustainability of the activities indicated above.

The **overall objective** of this presentation is to report research efforts geared at creating a prototype for conservation and sustainable use of biodiversity at risk of extinction through the creation of a Plant Gene Bank, including medicinal plants, fruit trees and other plant species of economic value, which would be of interest to the local population, and that would protect the watershed leading to increased and sustainable water yields, provide timber for construction, wood for fuel, and locally contribute to the reduction of the adverse effects of climate change and that can be replicated to areas with similar problems.

The specific objectives presently being pursued are many and include:

- Creation of a gene bank (both *in-situ* and *ex-situ*) for threatened, rare and endemic medicinal plants and other forest plants of economic value.
- Preparation of plant seedlings from indigenous plant species, for distribution to farmers and conduct of research on their husbandry.

- Provision of a framework for research by students and field days to farmers on the activities of the research park related to medicinal plants and other agricultural fruit crop production systems, sustainable soil and water management (Photo 1) and biodiversity conservation and mitigation of climate change impacts.

Photo 1: Left, *8-year old mahogany tree planted in 30x30x30 cm*^{3} *hole;* Right, *20 meters from plant at left, a soil profile exposing compact and dense horizons that constraint plant growth.*

Methodology: Study sites and characteristics

The 10-year old 4-ha Yongka Research Park is located at an altitude of 900 -1000 m on 2 to 30% slopes in Nkwen-Bamenda in the North West Region (Lat. 10^0 10" E and Long. 5^0 50" N) (Map 2), a part of the Western Highlands of Cameroon.

Geomorphologically, the area has a staircase configuration dominated by high lava plateaus surrounded by steep mountains and hills, below a plateau with a steep escarpment on basement rock and more or less dissected planation surfaces.

Geology: Five types of parent materials, which vary over very short distances exist in the area include: organic materials, alluvial/alluvio-colluvial deposits, colluvial deposits, mylonite and granite; granite is the most dominant.

Vegetation: With increasing population pressure much of the natural forest vegetation has degraded to secondary formations, or has been replaced by more or less wooded grasslands and agricultural land; this process is most advanced in areas with least relief.

Fig. 2: Map showing the study area in the west part of Bamenda town, NW Region.

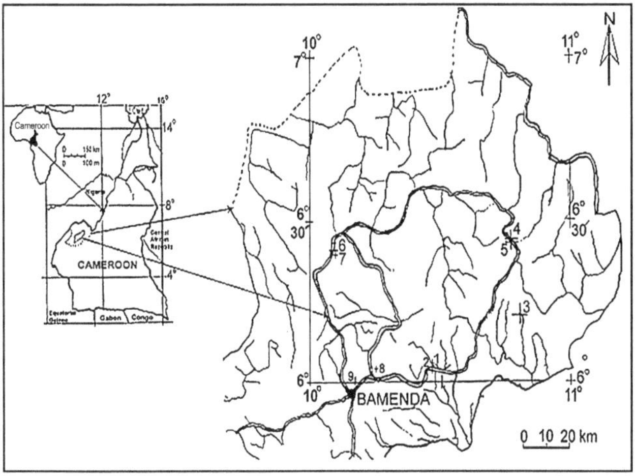

Detailed information on the vegetation structure of the North West Region has been given by Hawkins and Brunt (1965) and Kips *et al.* (1987).

Climate: The climate in the area is distinctly seasonal with one rainy season and one dry season (Hawkins and Brunt, 1965; Kips *et al.*, 1987). Rainfall at the research site averages about 2500 mm/year; mean annual temperatures average 19–22^{0}C.

Soils of the study area: Soils differ over short distances in such characteristics as texture, effective depth, gravel content, compactness and water infiltration rates aggravated by prior cattle rearing activities. Important differences in chemical characteristics and soil fertility exist and are related to parent material and stage of soil development and would impact on forest regeneration (Yerima *et al.*, 2003).

Research activities carried out at the Yongka Park

Many experiments which mirror the immediate objectives outlined above are currently being carried out at the Yongka Research Park.

Work includes zoning off the area to halt soil compaction due to overgrazing and avoid destruction of vegetation within the watershed; nursing of plant seedlings; digging of holes; collection of planting materials; identification of existing native plant species; establishment of fruit trees, medicinal plant and forest tree collections and experiments and their responses to different levels of manure applications; plant performance in response to soil profile modification techniques and nutrient status; experiments on water management techniques to improve root penetration; and collection of plant performance data and analysis.

Soil sampling and laboratory analysis

Photo 3: Left, *a very shallow soil profile that would limit root penetration;* Right, *water table obtains at a 1.5 m depth at lower part of park.*

Twelve soil profiles were sampled and analyzed and will provide information on soil variability and the effect of soil compactness on root penetration, water infiltration rate and potential plant performance (Photo 3), and would provide information for subsequent technology transfer of the research results to similar areas.

The major research areas

In-situ plant conservation

The local plants where they do not seriously interfere with the experiments are conserved (Photo 4). Other local plants undergoing extinction and are being preserved include: *Terminalia brownie* F, *Annona senegalensis*, Red oak (*Lophira lanceolata*), *Syzyglum guineense*, *Ficus sycomorus*, *Ficus natalensis* and *Ficus platyphylla*, among others. Patches of pure stands of natural vegetation have been conserved for training purposes.

Photo 4: Local forest ecosystems: Left, Terminalia glaucescens *Planch, loved by bees and sort after as a medicinal plant, in foreground; Right, grassland ecosystem with* Terminalia spp.*).*

Ex-situ plant collection

Many efforts are being made to collect plant species from all over the national territory as well as from similar ecological regions south of the Sahara where such plants could be at risk of extinction (Photo 5 a and b) and are preserved in the Yongka Park Gene Bank.

Photo 5 a: Left, *9-year old plant collection plot; Right, a mixt of Rupper (Hevea brasiliensis) and Njansang (Ricinodendron heudelotii) plants.*

Photo 5 b: Fast growing forest tree species.

Experiments on fruit trees

Many experiments on fruit trees are being conducted on the response of organic manure to plant performance, plant density/plant and hole size/plant performance. Experiments relating crop performance with hole-size and planting density in a RCBD with planting spacing of 5x4 m and hole sizes of 40x40x40 cm, 60x60x60 cm and 80x80x80x80 cm respectively are being carried out on many crops (Photo 6 a and b). Randomized Complete Block Designs (RCBD) was used for most experiments.

Other experiments on fruit trees include: Cashew nut (*Anacardium occidentale L.*), Jack fruit (*Artocarpus heterophyllus* Lam), Date palm (*Phoenix dactylifera L.*), Malay apple (*Eugenia malaccensis* L.), Guava (*Psidium guajava*), Grape fruits (*Citrus paradise* Maef), and Avocado (*Persea americana* Mill).

*Photo 6 a: Left, farmers visit passion fruit (*Passiflora edulis*) plot; Right, mango (*Mangifera indica *L) expt.*

*Photo 6 b: Bush plum (*Dacryodes edulis*) expt. All expts investigate effect of whole size/ plant performances, mulched to improve water retention/evaporation losses.*

Spices

As the forest ecosystem is disappearing many important tree spices are also being lost. Bush mango (*Irvingia gabonensis*) and other local tree plant species (Photo 7 a and b) are a case in point. Experiments here focus on manure/plant and hole size/plant performance and use of marcotted plant materials, which shorten the period the plant comes into production.

Photo 7 a: Left, Bush mango (Irvingia gabonensis) plant in hole mulched to conserve moisture; Right, 9-year old Njansa plants already bearing.

Photo 7 b: Below, marcotted young Njansa plants.

Collection of medicinal plants

With increasing demographic pressures most of the grasslands and gallery forests are being destroyed and replaced by farmland. This has led to the disappearance of indigenous medicinal plant species. Through poor harvesting practices of the barks of some of these plants (Photo 8 a and b) some of them dry up. The major medicinal plants under experimentation are: Pygeum (*Prunus africana)*, *Voacanga africana* Stapf, Iboga *(Tabernanthe iboga)*, Quinquina (F) (Cinchona) (*Cinchona officinalis*), and other yet unidentified tree species.

Photo 8 a: Terminalia glaucescens Planch plant, poorly harvested by local farmers.

Photo 8 b: Left, Voacanga africana; Middle, *Iboga* (Tabernanthe iboga; Right, Prunus africana).

Collection of export crops

Few primary and secondary school students in the area know physically what a cocoa tree (*Theobroma cacao* L.), tea plant (*Camellia sin*ensis L.), coffee robusta (*Coffea robusta*), Arabica coffee (*Coffea Arabica* L.), oil palm (*Elaeis guineensis* Jacq), and a rubber (*Hevea brasliensis*) plant looks like. One of the research problems is to collect all the economic export crops grown in the national territory (Photo 9 a and b) for studies and to observe their adaptability to the new environment. While it is clear most of the above plants are not undergoing extinction, the need to propose alternative economic crops to the farmers to reduce pressures on the endangered species is imperative. This will draw schools to visit the park for identification of the different tree species and hence expose them to the conservation of the local tree species at risk of extinction.

Photo 9 a: Left, *two tall green tea (*Camellia sinensis *L) plants;* Right, *tall rubber (*Hevea brasliensis*) plants in the foreground.*

*Photo 9 b: Marcotted cocoa (*Theobroma cocoa*) plant.*

Bio-fuels

Fuel wood and energy needs are often a major problem linked to the extinction of plant species. A substitute, which can fill the energy needs, would equally reduce the destruction of endangered plant species. *Jatropha curcas* L. is one of the promising bio-fuel plants. Experiments compare plant performance on stony soils to that on more fertile soils.

Experiments on forest trees

Adoption of fast growing environmentally friendly tree species for fuel and timber would reduce pressure on the endangered native tree species at risk of extinction. Experiments are being conducted on "exotic plant species" and native tree species that can serve multiple functions (fuel wood, timber, etc.). Experiments on the exotic plant species include pine and podocarpus tree species (Photo 10 a and b). A second set of plant experiments involves local forest tree species such as large diameter rattan (*Laccosperma secundiflorum* and *L. robustum* species), mahogany (*Swietenia macrophylla* King), iroko (*Milicia excelsa*) and plant species that have fuel wood, medicinal plants, timber, etc., attributes (Photo 11).

Photo 10 a: Left, *pine (Pius sp.)* Right, *Podocarpus (Podocarpus milanjianus) exp.: Both on effect of plant density on plant performance.*

Photo 10 b: 9-year old forest collection.

Photo 11: Left, *native tree species used for making walking canes;* Middle, *tall white mahogany plant in the foreground,* Right, *9-year old forest trees collection.*

Soil and water conservation practices

Plant performance is enhanced by adequate soil moisture and nutrient contents, good soil physical properties (low bulk density, absence of compact and dense layers, good aeration, etc.). The place of soils in the conservation of biodiversity and similar programs is little recognized. Experiments to enhance soil moisture content include use of mulching and micro-basin water catchment structures (Photo 12). Profile modification is carried out to reduce the soil bulk density, improve soil aeration and thus improve root penetration and plant performance.

Soil fertility practices

Plants need a fertile environment with appropriate nutrient contents to attain their optimum potential. At the Research Park only bio-fertilization (cow dung and pig manure) is used on plants. Decomposing grass mulch (Photo 12) constitutes the most important source of organic manures. Weeds or grass that is cleared is added under the plants to serve as mulch, which constitutes a constant source of nutrients on mineralization.

Photo 12: Left above, *Pig manure applied on mulched marcot bush plum seedling;* Right above, *Micro-basin water harvesting systems on the young mango* (Mangifera indica *L.) expt.* Below, *Bush plum* (Dacryodes edulis) *trees being mulched to reduce moisture evaporative losses.*

Plant nurseries, collection of seeds and plant cultural practices

One year old seedlings and marcotted and grafted seedlings for the experiments were either obtained from the ANAFOR forest nursery in the North West Region in Bamenda, bought from local nurseries in the urban and peri-urban areas, or nursed in the author's plant nursery at his residence or at the Yongka Research Park site itself (Photo 13).

Weeding is done at the end of May, August and November, yearly. Tree growth and performance parameters taken at 4-monthly intervals are plant height and plant girth (5 cm from the ground). Yield data will be taken when the fruits come into bearing.

Photo 13: Plant nursery showing seedlings and seed bed preparation.

Outreach activities

Research and outreach activities involving student study/internships tours at different levels of education, and farmer field days constitute viable pathways for up scaling this activity to similar problem areas thereby assisting in preserving plants at risk of extinction.

Through this culture, the conservation ethic is expected to be nurtured, grow and leading to the preservation of endangered plant species. Seven students from the University of Dschang, within a 2.5 month period of internship, were initiated into research and outreach activities involving the chain of activities from seed collection, through rising of seedlings in the nursery to the transplantation of seedlings in the field and husbandry for a wide range of plants within the 2009/2010 academic year. Another student is carrying out his thesis work on land evaluation of some fruit trees at the research park (Photo 14).

Photo 14: Top left, *Bio-science scientists from Univ. of Dschang visit the Prunus africana collection plot;* middle and right, *pictures, field work for students form Yaounde I Univ. (Bambili).* Bottom left and middle, *pictures, students form Univ. of Dschang during internship training;* right, *students describing soil profile. Students describing soil profile.*

Opportunities offered by the study

The study will contribute to the creation of a gene bank for medicinal plants and other plant species at risk of disappearance, provide information for formulating policies related to watershed management, scaling up of this activity to different areas, training on agro-forestry and allied activities and contribute to green spaces in the locality leading to the regulation of local climate, decreasing CO_2 levels in the atmosphere, and resolve some of the climate change problems associated with the "greenhouse effect".

The study has been going on for more than ten years. The multitude of specific objectives mirrors the complexity and the interwoven and multi-disciplinary nature of the research problem. The gene bank for collection of medicinal plants and fruit trees has implications for present and potential future health needs. Increased crop yields, adequate water supplies, green spaces in the peri-urban areas increase the quality of life of inhabitants in the adjoining urban and peri-urban areas and reduce pressure on endangered plant species.

Challenges

Challenges encountered in the execution of this study include: man-made wild fires and fires set to regenerate grass pastures for cattle and to catch rat moles; encroachment by man and cattle; high cost of land acquisition in the watershed areas in competition with alternate uses; fragmented and collectively owned land by family members requiring consent of all for its sale; many conflicts of land ownership as land is often sold to more than one buyer breeding litigations; high costs of labor for maintenance; compactness and denseness of the soils of granitic origin with low native fertility; destruction by termites of some plant species and difficulty in tillage of the termite mounts; and fragmentation of land making it difficult to propose appropriate and sustainable forest regeneration packages.

Conclusions and recommendations

Developing conservation and sustainable biodiversity practices in the area is a challenge and would require State/Municipal Councils to intervene by bearing some of the costs for establishment and upkeep of conservation sites.

The soil aspect in sustainable biodiversity conservation should be given high priority. Approaches to resolve the problem would be multi-faceted and include: sensitization of the local population on climate change impacts and conservation of endangered plant species; creation of a reward system for those actively engaged in this effort, while imposing dissuasive measures on non-sustainable practices; creation of protected areas with regulatory enforcement mechanisms to ensure sustainability; and institution of a legal framework addressing land use plans and the activities permissible within a given landscape.

References

Check, M., Onana, J. M. and B. J. Pollard, 2000. The plants of Mount Oku and the Ijim Ridge, Cameroon, A Conservation Checklist. RBG, Kew, UK. 211 pp.

Conrade, G., 1974. Commentaire des cartes. Atlas Regional. Ouest 1. ORSTOM, Yaounde.

Echholm, E., Foley, G., Barnard, G. and L. Timberlake, 1984. Fuel wood: The energy crisis that won't go away. London: Int. Inst. for Environment and Development.

FAO, 1985. Committee on Forest Development in the Tropics. Tropical Forests Action Plan, FAO-Rome.

FAO, 1990. Guidelines for soil profile descriptions. 3rd ed. FAO-ISRIC, Rome.

Harvey, Y., Pollard, B. J., Darbyshire, I., Onana, J. M. and M. Cheek, 2004. The plants of Bali Ngemba Forest Reserve, Cameroon. Royal Botanic Gardens, Kew, TW9 3AB, UK. 154 pp.

Hawkins, P. and M. Brunt, 1965. The soils and ecology of West Cameroon. Vols. I and II. FAO Expanded Programme of Technical Assistance No. 2038. FAO, Rome. 516 pp.

Institute of Biodiversity Conservation (IBC) of Ethiopia. 2009. Project report. Addis Ababa, Ethiopia. http://www.ibc.gov.et.

Kips, Ph. A., Faure, P., Awah, E. T., Kuoh, H. M., Sayol, R. and F. Tchuenteu, 1987. Soils, land use and land evaluation of the North-West-Province, Cameroon. FAO-UNDP Project CMR/83/004. IRA/CNS-Ekona Station, PMB 25, Buea.

UNDP, 1994. Elaboration of the National Environmental Management Plan for Cameroon: Situation Analysis of the Environment in the NWP. UNDP Project CMR/92/88. Yaounde.

Yerima, B. P. K. and E. van Ranst, 2005a. Major soil classification systems used in the tropics: Soils of Cameroon. Trafford Publishing. Victoria BC, Canada. 312 pp.

Yerima, B. P. K. and E. van Ranst, 2005b. Introduction to soil science: Soils of the tropics. Trafford Publishing. Victoria BC, Canada. 440 pp.

Yerima, B. P. K., Wilding, L. P. and L. R. Hossner, 2003. Review of environmental factors and evaluation of soil properties of North Cameroon, Vertisols and Alfisols as a guide to forest regeneration. Sci. Agron. and Dev. 3(1): 26–36.

Studies in Sub-Saharan Africa

Edited by Support Africa International/Support Africa Stiftung
Represented by Baldur Pfeiffer

Vol. 1 Uche C. Amalu / Franz-Theo Gottwald (eds.): Studies of Sustainable Agriculture and Animal Science in sub-Saharan Africa. 2004.

Vol. 2 Franz-Theo Gottwald / Susan Keino / Timothy Rotimi Fayeye (eds.): Fostering Subsistence Agriculture, Food Supplies and Health in Sub-Saharan Africa. 2007.

Vol. 3 George Ouma / Franz-Theo Gottwald / Isabel Boergen (eds.): Agrarian Science for Sustainable Resource Management in Sub-Saharan Africa. 2009.

Vol. 4 Frackson Lameck Mkandawire / Mmaduabuchukwu Mkpado (eds.): Low-Input Agricultural Technologies for Sub-Saharan Africa. 2011.

Vol. 5 Sunday Paul Bako / Frank Olwari (eds.): Promoting Non-Animal Protein Sources in Sub-Saharan Africa. An Interdisciplinary Study. 2011.

Vol. 6 Chukwuemeka Christopher Opara / Bernard Palmer Kfuban Yerima (eds.): The Impact of Climate Change on Sub-Sahara Africa. Case Studies in Cameroon, Nigeria and Uganda. 2014.

www.peterlang.com

www.ingramcontent.com/pod-product-compliance
Ingram Content Group UK Ltd.
Pitfield, Milton Keynes, MK11 3LW, UK
UKHW020121290926
15471UKWH00003B/205

9 783631 653470